EPA United States
Environmental Protection
Agency

Office of Water

Office of Research and Development

September 2015

EPA-800-R-15-002

National Ecosystem Services Classification System (NESCS): Framework Design and Policy Application

Final Report

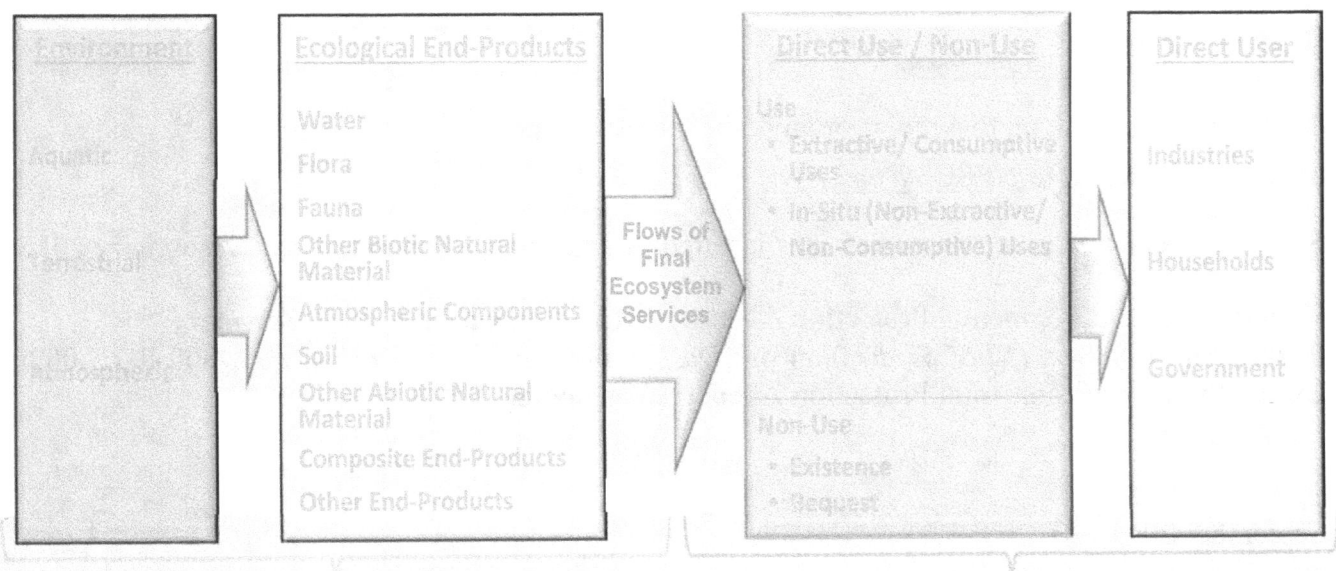

NESCS Four-Group Classification

ACKNOWLEDGEMENTS

The authors thank Jennifer Richkus, Jennifer Phelan, Robert Truesdale, Mary Barber, David Bellard, and others from RTI International for providing feedback and research support during the development of this report. The early leadership of former EPA employee John Powers proved instrumental in launching this effort. The authors thank Amanda Nahlik, Tony Olsen, Kevin Summers, Kathryn Saterson, Randy Bruins, Christine Davis, Bryan Hubbell, Julie Hewitt, Ashley Allen, Todd Doley, Karen Milam, David Simpson, and others at EPA for their discussion and feedback on earlier versions of this document. In addition, the authors thank V. Kerry Smith, Neville D. Crossman, and Brendan Fisher for review comments. Finally, the authors would like to thank participants of the two NESCS Workshops held in 2012 and 2013, as well as participants of an ACES session in 2014. Any factual or attribution errors are the responsibility of the authors alone.

ADDITIONAL INFORMATION

This document was developed under U.S. EPA Contract EP-W-11-029 with RTI International (Paramita Sinha and George Van Houtven), in collaboration with the ORISE Participant Program between U.S. EPA and U.S. DOE (Charles R. Rhodes), under the direction of Joel Corona and Dixon Landers, U.S. EPA, Office of Water and Office of Research and Development, respectively. Peer review for this report was conducted under U.S. EPA Contract EP-C-12-045 with Versar, Inc. (David Bottimore).

This report may not necessarily reflect the views of U.S. EPA and no official endorsement should be inferred.

To provide feedback on this report or any other aspect of the NESCS approach, please send comments by email to NESCS@epa.gov.

Small cosmetic and grammar corrections last updated within this report (v1.1): 29 February 2016.

United States Environmental Protection Agency. 2015. *National Ecosystem Services Classification System (NESCS): Framework Design and Policy Application.* EPA-800-R-15-002. United States Environmental Protection Agency, Washington, DC.

EPA-800-R-15-002
September 2015

CONTENTS

Appendices

   A   Mathematical Representation of the Conceptual Model
   B   Expanded Conceptual Framework for Ecosystem Services Analysis

LIST OF FIGURES

LIST OF TABLES

ACRONYMS AND ABBREVIATIONS

BEA	Bureau of Economic Analysis
CAA	Clean Air Act
CAFO	concentrated animal feeding operation
CBA	cost-benefit analysis
CEA	cost-effectiveness analysis
CICES	Common International Classification of Ecosystem Services
CO_2	carbon dioxide
COICOP	Classification of Individual Consumption by Purpose
CPC	Central Products Classification
ECPC	Economic Classification Policy Committee
EPA	U.S. Environmental Protection Agency
ESI	Ecosystem Services Index
F&T	Fisher and Turner (2008)
FCA	Full Cost Accounting
FEGS	Final Ecosystem Goods and Services
FEGS-CS	Final Ecosystem Goods and Services Classification System
FFES	flows of final ecosystem services
FOEN	Federal Office for the Environment
GDP	gross domestic product
GIS	geographic information system
GNP	gross national product
I-O	input-output
IPIECA	International Petroleum Industry Environmental Conservation Association
ISIC	International Standard Industrial Classification
ISIC V4	International Standard Industrial Classification of All Economic Activities
MA	Millennium Ecosystem Assessment
NAFTA	North American Free Trade Agreement
NAICS	North American Industry Classification System
NAPCS	North American Product Classification System
NEA	National Economic Accounts
NESCS	National Ecosystem Services Classification System
NESCS-D	National Ecosystem Services Classification System (Demand-Side)
NESCS-S	National Ecosystem Services Classification System (Supply-Side)
NESP	National Ecosystem Services Partnership

NIPA	National Income and Product Accounts
NOx	nitrogen oxides
NRC	National Research Council
SAB	Science Advisory Board
SEEA	System of Integrated Environmental and Economic Accounts
SIC	Standard Industrial Classification
SNA	System of National Accounts
SOx	sulfur oxides
TEV	Total Economic Value
UN	United Nations
USDA	U.S. Department of Agriculture
WAVES	Wealth Accounting and the Valuation of Ecosystem Services
WTP	willingness to pay

ABSTRACT

Understanding the ways in which ecosystems provide flows of "services" to humans is critical for decision making in many contexts; however, the linkages between natural and human systems are complex and multifaceted. A well-defined framework for classifying ecosystem services is essential for systematically identifying and tracing these linkages. The purpose of this report is to describe the National Ecosystem Services Classification System (NESCS), which is designed to address these needs.

The main objective of NESCS is to provide a framework that will aid in analyzing the human welfare impacts of policy-induced changes to ecosystems. In particular, it is intended to support different types of policy impact analyses, such as cost-benefit analysis of environmental regulations. Measuring the welfare impacts of alternative environmental policy or natural resource management scenarios typically entails three main steps: identifying, quantifying, and (as feasible) valuing changes in ecosystems and their contributions to human well-being. NESCS is primarily designed to support the first step—identifying ecosystem service changes—and thus provides a foundation for the subsequent steps of quantification and valuation. It is not an accounting system, but it is designed to support comprehensive and systematic accounting of changes in ecosystem services. NESCS could also potentially be used to support analysis of other policies that could result in changes to ecosystems such as housing, transportation, and tax policies.

The conceptual framework for NESCS was developed by applying the principles underlying existing classification and accounting systems for *economic* goods and services, such as the North American Industry Classification System (NAICS), the North American Product Classification System (NAPCS), and the National Income and Product Accounts (NIPA). As others have done using these economic principles (e.g., Boyd and Banzhaf, 2007), NESCS draws a key distinction between intermediate and final services. For both economic and environmental accounting, this distinction is essential to avoid double counting services. Consequently, the NESCS focuses on **flows of final ecosystem services (FFES)**, which it defines as the *direct* contributions made by nature to human production processes or to human well-being.

In NESCS, FFES are identified by linking the ecological systems that supply final ecosystem services with the human systems that demand them. Human systems include both the market-sector producers who directly use the outputs of nature to produce economic goods and services, and the non-market-sector households who directly use or appreciate the outputs of nature to "produce" human well-being. They can also include public sector entities that directly use the outputs of nature to produce public goods and services.

To uniquely identify and classify FFES, the NESCS structure consists of four classification groups:

1. environmental classes, which are spatial units with similar biophysical characteristics, that are located on or near the Earth's surface, and that contain or produce "end-products" (e.g., aquatic, terrestrial, atmospheric);

2. classes of ecological end-products, which are the biophysical components of nature directly used or appreciated by humans;

3. classes of direct human uses (extractive or in situ) or non-use appreciation of end-products; and

4. classes of direct human users of end-products.

The first two groups represent the "supply-side" components of ecosystem services production (NESCS-S) and the last two groups represent the "demand-side" (NESCS-D). Each unique combination of classes (or subclasses) from these four groups defines a distinct FFES category. As such, each one represents a unique potential pathway for linking changes in ecosystems to changes in human welfare.

To demonstrate NESCS, we provide two general examples illustrating how the conceptual framework and classification system can be used to identify pathways linking specific policy actions to human welfare changes. The first example examines a hypothetical policy to reduce atmospheric deposition of nitrogen and sulfur. It identifies and describes multiple FFES pathways that link changes in deposition to uses and users of the impacted forest and aquatic ecosystems. The second example focuses on a hypothetical policy requiring wetlands restoration. The example identifies specific ecological end-products that are affected by wetland restoration and the corresponding FFES that are provided to producers and households.

EXECUTIVE SUMMARY

ES.1   Introduction

Ecosystems provide flows of services to humans and thus contribute to human welfare in numerous and often complex ways. Identifying and tracing these linkages between natural and human systems are crucial for supporting decision-making in many contexts. Specifically, these linkages are important for analyzing the human welfare impacts of changes to ecosystems due to policy or management actions. A growing literature in ecosystem services research has focused on defining and grouping these linkages; however, the interdisciplinary nature of the topic and the complexity of these linkages make classifying ecosystem services a challenging task. Among other things, it requires a common understanding between natural scientists and social scientists of ecosystem service concepts and definitions.

The primary objective of this report[1] is to provide a classification system, which we refer to as the National Ecosystem Services Classification System (NESCS) that will aid in analyzing the human welfare impacts of policy-induced changes to ecosystems. In particular, the goal of NESCS is to support different types of "marginal" analysis, such as cost-benefit analysis, which focus on changes from baseline conditions. Measuring the welfare impacts of environmental policy changes typically entails three main steps: identifying, quantifying, and (as feasible) valuing changes in ecosystems and their contributions to human welfare. NESCS is primarily designed to support the first step—identifying ecosystem service changes and thus provides a foundation for conducting the subsequent steps of quantification and valuation. Although not the primary focus, NESCS also supports comprehensive and systematic accounting of changes in ecosystem services. NESCS could also potentially be used to support analysis of other policies (e.g., housing, transportation, tax policies) that could result in changes to ecosystems.

In designing NESCS, we have adapted concepts, principles, and methods from several streams of literature. First, we attempt to incorporate broad underlying characteristics and best practices of classification systems. Second, we draw from previous literature on classification approaches for ecosystem services. Third, we draw from widely accepted concepts for classifying and accounting for flows of services in the economic context and adapt them to the context of ecosystem services.

The primary goal of supporting marginal analysis defines the key requirements for NESCS. To support marginal analysis, it is important to have a standardized, comprehensive

---

[1]   Key terms used throughout this report are defined in a Glossary at the end of the report.

system that will allow for systematic linkages to be drawn between natural and human systems. It is important to ensure that the classification system allows all potential impacts from a policy change to be accounted for. At the same time, it is important to avoid double counting impacts. NESCS is based on a conceptual framework that provides a way to systematically link *ecological systems* that produce ecosystem services and *human systems* that directly use or appreciate these services (i.e., market production systems and households). By definition, ecosystem services only exist when they contribute to human well-being. The NESCS structure defines categories and numeric codes that are designed to help identify flows of services from ecosystems to human beings in a comprehensive and mutually exclusive way. This executive summary provides an overview of the report, describing the key topics addressed in each of the six chapters.

ES.2 Review of Literature on Classifying Ecosystem Services and Implications for NESCS

Since the publication of the seminal work, *Nature's Services* (Daily, 1997), a large literature has evolved proposing alternative definitions and classification approaches for ecosystem services. Our review of this literature specifically includes studies by de Groot et al. (2002), the Millennium Ecosystem Assessment (MA) (2005), Wallace (2007), Boyd and Banzhaf (2007), Fisher and Turner (2008), Haines-Young and Potschin (2010a, 2010b, 2013), Staub et al. (2011), and Landers and Nahlik (2013). Although the MA (2005) classification—which divides ecosystem services into provisioning, cultural, regulating, and supporting service categories—has been most widely cited, other studies in our review propose alternative systems, including the Common International Classification of Ecosystem Services (CICES; Haines-Young and Potschin, 2010a, 2010b, 2013) and the Final Ecosystem Goods and Services Classification System (FEGS-CS) (Landers and Nahlik, 2013).

Although the fundamental common purpose of this literature is to identify and describe the various ways in which ecosystems support human welfare, our review indicates there are wide differences in policy and management objectives, specific definitions of ecosystems services, and criteria for grouping services. Although there is general agreement that (1) ecosystems are natural assets that support human welfare in many ways and (2) this support of human well-being is fundamental to the concept of "ecosystem services," there is continued disagreement about where exactly ecosystem services occur along the continuum between ecosystems and human welfare. In particular, there is disagreement regarding the difference between ecosystem processes, functions, services, and benefits.

To address the question of where ecosystem services lie along the continuum, Boyd and Banzhaf (2007) introduce and focus on the concept of **final ecosystem services**. As they define them, final ecosystem services occur at the point of hand-off between natural systems (ecosystems) and human systems (producers and households). In contrast, **intermediate ecosystem services** are inputs to the natural processes that ultimately produce final ecosystem services. For example water purification is important for sustaining fish populations, but fish contribute directly to commercial fishing. As such, their value is embedded within the value of final ecosystem services. Distinguishing between final and intermediate ecosystem services is essential to avoid double counting their values.

Realizing this important distinction, NESCS was designed to specifically focus on and classify final ecosystem services.

ES.3 Review of Economic Classification and Accounting Systems and Implications for NESCS

To develop a classification system for *ecosystem* services, we applied concepts and methods underlying existing classification and accounting systems for *economic* goods and services. In economics literature, in contrast to goods, which can be treated as "stocks," services are typically viewed as "flows" from the provider to the consumer and are measured over time. In the United States, the two main classification systems are the North American Industry Classification System (NAICS) and the North American Product Classification System (NAPCS). NAICS focuses on how and by whom goods and services are produced; therefore, it can be interpreted as a "supply-side" system. NAPCS, on the other hand, focuses on how and by whom goods and services are used. It can be interpreted as a "demand-side" system. Both of these classification approaches were primarily designed to support the development of National Income and Product Accounts (NIPA). NIPA are used to (1) trace the flow of *intermediate* goods and services between production sectors in the economy, and (2) estimate the value and composition of *final* goods and services sold to consumers. This helps avoid double counting their values which is important to ensure valuations and trade-off analyses are valid.

Important parallels can be drawn between economic and ecosystem services, but there are also important dissimilarities, reflecting unique characteristics of ecosystem services. First, in contrast to economic services, ecosystem services are typically non-market in nature—that is, they are not sold in markets and thus there are fewer observable transactions or prices. Second, unlike most economic services, ecosystem services often have "non-rival" characteristics. In other words, enjoyment by one user does not diminish simultaneous enjoyment by other users. Third, whereas final economic services are only sold to end users (households), final ecosystem services, which occur at the "point of direct hand-off" from natural systems to human systems, flow both to producers of economic goods and directly to households and to governments.

ES.4 NESCS Conceptual Framework, Classification Structure, and Coding System

NESCS adapts and modifies the economic principles described in ES.3 to reflect unique characteristics of ecosystem services. Since services are viewed as flows from providers to consumers, NESCS identifies and distinguishes between the producers (i.e., "supply-side") and users (i.e., "demand-side") of the service. However, in NESCS, the supply-side refers to the natural systems that provide ecosystem services and the demand-side refers to the human systems that directly use or appreciate them. NESCS extends the NAICS/NAPCS framework noted in Figure ES-1 to trace the flow of ecosystem services from natural systems to human systems.

NESCS focuses on **flows of final ecosystem services (FFES),** which it defines as the direct contributions made by nature to human production processes or to human well-being.[2] The linkage between the ecological systems that supply final ecosystem services with the human systems (market production sectors and households) that directly use or appreciate these services identifies FFES.

Figure ES-1. Conceptual Framework Including Flows of Final Ecosystem Services (FFES) as Inputs to Human Systems

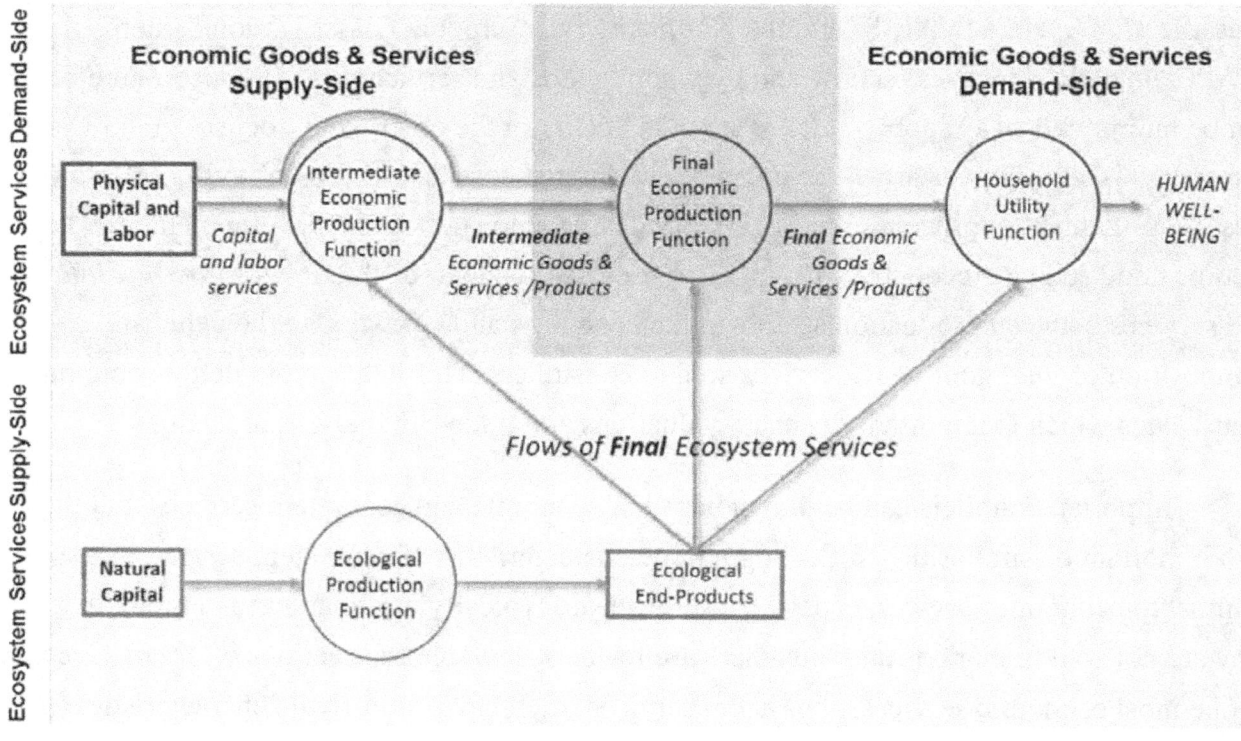

---

[2] It is important to note that *flows of final ecosystem* **goods** are not included or defined in this framework. The main reason for this exclusion is that the process of transferring physical ecosystem products from nature to humans, which is necessary to generate flows of goods, typically requires *human* inputs. For example, agricultural and forest products that are sold in the market require human inputs to harvest and process. Thus, these are considered *economic* goods and not flows of final *ecosystem* goods in NESCS.

Table ES-1. NESCS Example

	NESCS–S		NESCS–D	
Group	Environment	End-Product	Direct Use/Non-Use	Direct User
Definition	Spatial units with similar biophysical characteristics, that are located on or near the Earth's surface, and that contain or produce "end-products"	Biophysical components of nature that are directly used or appreciated by humans	Different ways in which end-products are used or appreciated by humans	Entities that directly use or appreciate the end-products
Hierarchy and Coding System **NESCS Code for FFES*: WW. .YYYY.ZZZZZZ**				
Class	W	WW.X	WW.XX.Y	WW.XX.YYYY.Z
Subclass	WW	WW.XX	WW.XX.YY	WW.XX.YYYY.ZZZ
Detail			WW.XX.YYYY	WW.XX.YYYY.ZZZZZZZ
Example 1: Water in the ocean being used as a medium for freight transportation NESCS Code for FFES: 15.12.1202.1483111				
Class	Aquatic: **1**	Water: 1	Direct Use: **1**	Industry: **1**
Subclass	Open Ocean and Seas: **15**	Liquid Water: **12**	In-Situ Use: **12**	Transportation and Warehousing: **148**
Detail			Transportation medium: **1202**	Deep Sea Freight Transportation: **1483111**
Example 2: Water in rivers being extracted for household gardening purposes NESCS Code for FFES: 11.12.1105.201				
Class	Aquatic: **1**	Water: 1	Direct Use: **1**	Households: **2**
Subclass	Rivers and Streams: **11**	Liquid Water: **12**	Extractive Use: **11**	Households: **201**
Detail			Support of plant or animal cultivation: **1105**	

* Note that this 15-digit code is the most disaggregated level of representation. Different levels of aggregation can be used depending on the context (See Examples 1 and 2 for different levels of aggregation for users)

The NESCS structure (represented in Table ES-1) consists of four groups:

1. Environment: These are defined as spatial units, with similar biophysical characteristics, that are located on or near the Earth's surface and that contain or produce "end-products" covers the earth's natural systems and can be interpreted as producers of **ecological end-products**. The categories for this system are obtained from Landers and Nahlik (2013).

2. End-Products: These are defined as biophysical components of nature that are directly used or appreciated by humans.[3]

3. Direct Use/Non-Use: This group defines different ways in which end-products are directly used or appreciated by humans in a way that is consistent with common valuation frameworks used by economists, such as the Total Economic Value (TEV) framework.

---

[3] This definition is very similar to definition used in Landers and Nahlik (2013) and Boyd and Banzhaf (2007).

4. Direct Users: This group represents the sectors that directly use or appreciate end-products. We follow established classification structures adopted by the U.S. Census Bureau and the United Nations.

This four-group classification structure (with examples of classes and subclasses within them) and the flows between them are represented in Figure ES-2. The first two groups pertain to the natural systems that "produce" FFES and can be interpreted as the **supply-side classification** (NESCS-S). The last two groups pertain to the human systems that appreciate or directly use FFES and can be interpreted as the **demand-side classification** (NESCS-D). Within each of these four groups, NESCS adopts a nested hierarchical structure so that each group can be represented at multiple levels of aggregation or detail.

Figure ES-2. Four-Group NESCS Structure

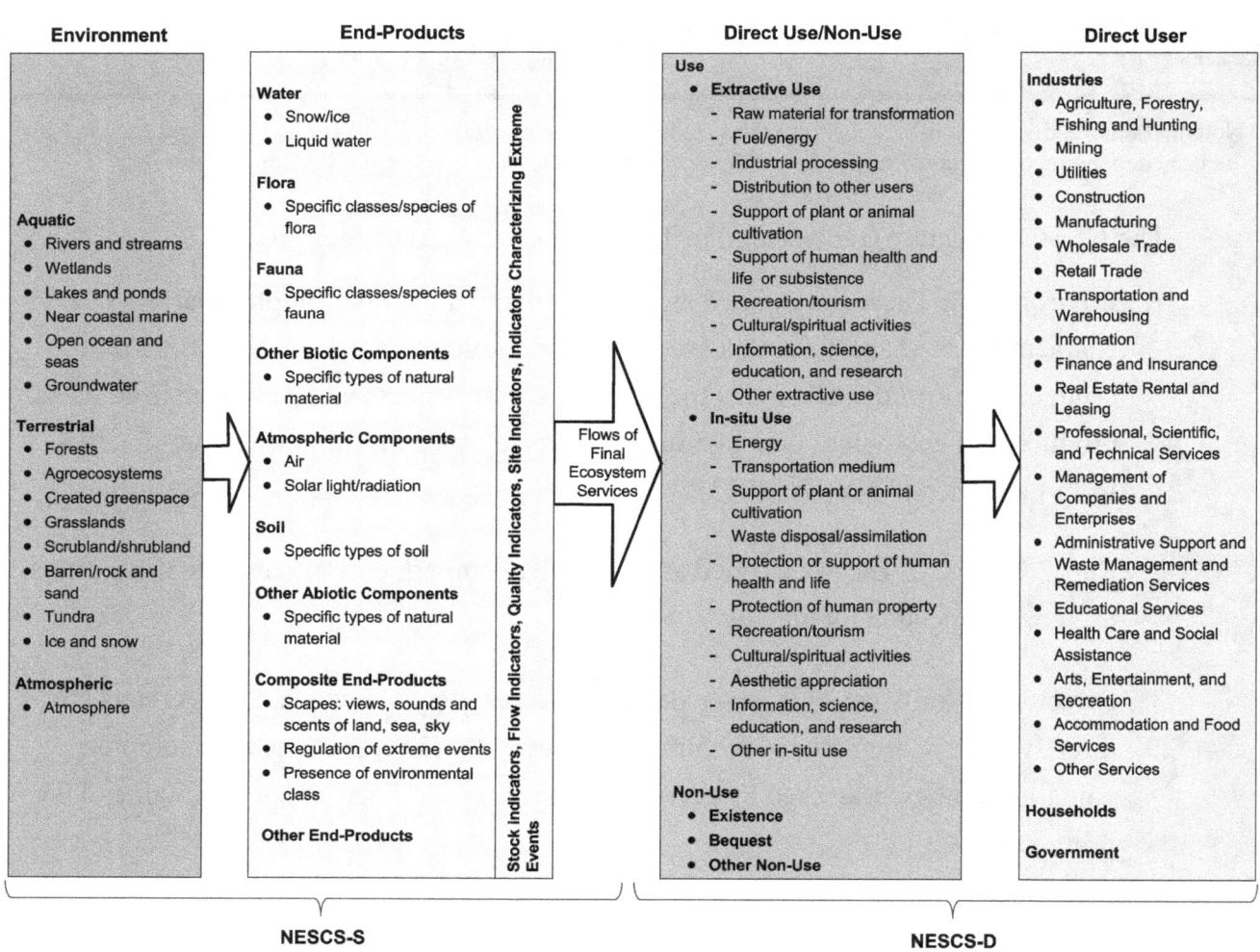

Each unique combination of individual elements from each of the four groups defines a separate FFES. In other words, it represents a unique potential pathway through which changes in ecosystems may affect human welfare. The ability to define different combinations allows the NESCS structure to be flexible and comprehensive. For example, it recognizes that the same ecological end-product category may be used in multiple ways (e.g., water can be used to support human life as drinking water and as an energy source through hydropower production). It also recognizes that a single use category can be linked to multiple different user categories. For example, water use to support plant cultivation is relevant for both the agricultural sector and households (e.g., for lawn watering).

In addition to the flexible classification structure, NESCS provides a coding system that allows for a numeric representation of the system's structure. The categories in each of the four groups are assigned numeric codes. Each unique FFES can be easily be referenced and identified by a detailed NESCS code that could potentially use up to 15 digits. Box ES-1 summarizes the primary NESCS concepts and definitions introduced in this section.

ES.5 Applying NESCS to Policy Analysis

In Section ES.4, we summarize the NESCS framework, classification structure, and coding system. In Table ES-2, we summarize how the NESCS can be applied to identify and reference unique FFES pathways linking changes in policy and/or management action to changes in ecosystems to changes in human welfare.

To demonstrate how NESCS can be applied to support policy analysis, we provide two very different hypothetical policy applications. The first examines a policy that reduces atmospheric deposition of acidifying compounds, such as nitrogen oxides (NO_x) and sulfur oxides (SO_x). These changes, which affect the quality of terrestrial and aquatic environments, are assumed to occur on a national or large regional scale. This first policy application identifies and describes multiple FFES pathways that link changes in acid and nutrient deposition to specific uses and users of the affected forest and aquatic ecosystems.

The second application focuses on a hypothetical policy requiring wetland restoration. In this case, the direct policy impact can be characterized as a change in the quantity of natural capital in an environmental class—wetlands. These changes are assumed to occur on a local or small regional scale. The example identifies a range of resulting FFES that are provided to producers and to households.

ES.6 Conclusions

In summary, Box ES-2 describes the key features of NESCS, including what it does and does not do. The main objective of NESCS is to support the analysis of various policy changes. Additional applications of the system will be needed to evaluate and further verify its usefulness for this purpose and to determine whether and how the system can best be modified to address future needs. For example, although not specifically intended for other uses, the NESCS framework and classification structure may prove useful for certain green accounting applications. Because NESCS draws from macro-accounting structures such as NIPA, it might prove to be a useful tool for green-gross domestic product accounting. It may also help with environmental accounting systems being adopted at a more micro-level by private and local public sector organizations. NESCS could also potentially be used to support analysis of other policies (e.g., housing, transportation, tax policies) that could result in changes to ecosystems.

Although NESCS provides a detailed structure for classifying FFES, certain questions and challenges remain for ecosystem service classification. Key among these issues is how to address ecosystems that are heavily managed by humans. As a simplifying assumption, the NESCS conceptual framework assumes there is a clear division between natural systems and human systems. In practice, however, some degree of human management is present in most ecosystems. Additional investigation and applications will be needed to determine how to best address these "gray" areas, where separating natural and human systems is inherently more complicated.

Table ES-2. How to Apply the NESCS Structure to Identify and Represent Unique FFES Pathways for Policy Analysis

How to...	NESCS Tools
...describe FFES pathways that may potentially be impacted by a policy change in a systematic and consistent manner?	Use NESCS conceptual framework (Figure 4-3) as guide
...identify unique FFES pathways?	
1. Identify the environmental classes/subclasses and corresponding end-product classes/subclasses that are likely to be impacted based on region-specific scientific evidence and information.	• Classification of Environment (Table 4-2) • Classification of End-Products (Table 4-3) • End-products in Each Environmental Class (Table 4-5)
2. Identify the specific combinations of end-products and direct uses/non-uses that are likely to be impacted	• NESCS Table Linking End-Products and Direct Uses/Non-Uses (Table 4-9) • Classification of Direct Use/Non-Use (Table 4-6)
3. Identify relevant user categories that directly use the end-products that are likely to be impacted	• NESCS Table Linking Direct Uses/Non-Uses with Users (Table 4-8) • Classification of Direct User (Table 4-7)
...reference and illustrate FFES pathways in a readily understandable manner?	
1. Diagrammatically	Fill in NESCS conceptual framework with categories identified (See Figures 5-1 through 5-5 as examples)
2. Numerically	Use NESCS 15-digit coding system (Tables 4-1, 4-2, 4-3, 4-5, 4-6, 4-7, 4-8, and 4-9)
...provide a structure that can be used to store values obtained from elsewhere?	
1. Use tables that link each of the four groups to organize, store, and present values (monetized or otherwise) that are obtained from other sources	• End-products in Each Environmental Class (Table 4-5) • NESCS Table Linking End-Products and Direct Uses/Non-Uses (Table 4-9) • NESCS Table Linking Direct Uses/Non-Uses with Users (Table 4-8)

Box ES-2. Key Features of the National Ecosystem Services Classification System (NESCS)

NESCS supports policy analysis in the following main ways:

(1) **Provides consistency and clarity in defining final ecosystem services:** NESCS provides an explicit conceptual framework for defining flows of final ecosystem services (FFES) from natural systems to human beings. It does this by clearly distinguishing FFES from (a) the ecological production functions/processes that produce them; and (b) the goods and services produced by human beings (particularly those requiring natural inputs, such as crops that require water and soil fertility).

(2) **Is designed to avoid double counting of ecosystem services:**[1] NESCS does this by (a) distinguishing between intermediate ecological production functions/processes and final ecosystem services; (b) striving to define mutually exclusive use categories; and (c) distinguishing between direct (e.g., fruit growers) and indirect users (e.g., households that consume fruit from growers).

(3) **Is designed to be flexible and comprehensive:** NESCS provides a broad and flexible modular structure intended to be as comprehensive as possible in capturing potential pathways from ecosystems to human beings and thus avoid omission of ecosystem service categories (including categories that may become important in the future).

(4) **Helps reference and illustrate ecosystem service pathways:** NESCS categories and codes are designed to help a policy analyst identify and reference flows from ecosystems to human beings in a consistent way. The NESCS framework can also be used to represent pathways diagrammatically and in a readily understandable manner.

(5) **Provides tools and structure for storing values obtained from elsewhere:** NESCS provides a structure and a set of tools/tables that can be used to organize, store, and present values (monetized or otherwise) that are obtained from other sources such as the non-market valuation literature.

NESCS can also be characterized in part by what it does <u>not</u> do or include:

(1) **Does <u>not</u> conduct valuation of ecosystem services**: NESCS does not attempt to conduct quantification or valuation. The goal is to support *identification* of pathways between ecological and human systems, which can then be used as a basis or starting point for quantification or valuation.

(2) **Is <u>not</u> a macro-accounting system:** NESCS draws from certain elements of macro-accounting structures such as the North American Industry Classification System (NAICS), the North American Product Classification System (NAPCS), and the National Income and Product Accounts (NIPA). It might also prove to be a useful tool for green-gross domestic product accounting, although this is not the fundamental purpose of NESCS.

(3) **Does <u>not</u> define or categorize feedbacks from human systems to natural systems:** NESCS defines flows *from* natural systems *to* human systems and not feedback effects from human to natural systems. It is important to note that this is by design and does not limit consideration of these dynamic and feedback effects when quantifying and valuing ecological benefits. Feedbacks may generate more flows through the NESCS system and require that more of the existing FFES pathways be considered. However, considering these feedbacks does not imply that new pathways will need to be *defined* and *classified*.

(4) **Does <u>not</u> include a separate category for health effects, but defines numerous pathways that include human health and safety:** To be comprehensive, NESCS is designed to account for numerous, complex connections between the environment and human health. Rather than defining a separate ecosystem service category that exclusively addresses health effects, it defines a multitude of pathways that include human health or safety as key components.

---

[1]. There will inevitably be "gray" areas where overlaps may exist; however, NESCS is intended to minimize those overlaps.

SECTION 1
INTRODUCTION

1.1 Overview

There is emerging consensus that understanding how ecosystems contribute to human welfare is critical to public- and private-sector decision making. People derive benefits from ecosystems in a myriad of ways or, put in a different way, ecosystems provide flows of "services" to people in numerous ways. The linkages between natural systems and human systems are complex so that identifying and tracing pathways between them can be challenging. These linkages are the main focus of the literature on ecosystem services classification that has gained momentum since the seminal work of Daily (1997). Classifying **ecosystem services** is inherently interdisciplinary and requires a common understanding of concepts and methods between natural scientists (e.g., ecologists) and social scientists (e.g., economists). A review of this literature reveals that although there is consensus on the notion that ecosystems are natural assets that support human welfare, there is disagreement on where ecosystem services occur along the continuum between ecosystems and human welfare. A consistent definition and classification system is critical for research and efficient decision making.

The purpose of this report[4] is to describe a classification system for ecosystem services—the National Ecosystem Services Classification System (NESCS)—that is based on a consistent conceptual framework and definition. The primary goal of NESCS is to support analysis of the human welfare impacts of environmental and natural resource management policies. It is important to note that analysis of policies involves evaluations of *changes* to the system rather than evaluating the status of the *total* system. In other words, the goal of NESCS is to support different types of "marginal" analysis.[5] For example, it should be particularly helpful for conducting cost-benefit analyses (CBA) of environmental and natural resource management policies. In CBA, the main objective is to measure changes in human welfare by estimating and comparing the benefits and costs of policies, both measured in monetary terms. The classification system should also provide a framework for comparing the cost-effectiveness or distributional impacts of alternative policies. In a cost-effectiveness analysis, alternative policy outcomes may be evaluated by comparing non-monetary measures of ecosystem service

---

4   Key terms used throughout this report are defined in a Glossary at the end of the report.
5   Policies that are relevant in this context are typically those that cause *changes* to ecosystems that are small *relative* to the *total* value of ecosystems (e.g., implementing or changing water quality standards, changing emissions standards for a source category). The term "scenario analysis" is also used (e.g., National Ecosystem Services Partnership, 2014) for environmental policy analysis since several alternatives or scenarios are evaluated during the course of decision making. Although broader in scope, these types of analysis may also be supported by the NESCS framework.

improvements, which serve as effectiveness indicators, and monetary measures of costs. In a distributional analysis, the impacts of ecosystem service improvements (and costs) on different subpopulations can be evaluated and compared. NESCS could also potentially be used to support other types of marginal analysis such as analyzing impacts of other policies (e.g., housing, transportation, tax policies) that could also result in changes to ecosystems. In addition, although it is not the primary objective of the classification system, we expect that it will provide a useful framework for conducting environmental or "green" gross domestic product (GDP) accounting, at both a microeconomic and a macroeconomic level.[6]

Analyzing the human welfare impacts (benefits) of an environmental policy typically entails identifying, quantifying, and, in many cases, valuing changes in ecosystems and their contributions to human welfare (EPA, 2009). The Science Advisory Board (SAB) report stresses on the importance of this "identification" step in valuation even when data issues limit monetization of impacts (EPA, 2009). One of the findings of the report is that historically, policy analysis has tended to focus only on ecosystem services for which economic benefits are easily measurable but this "can diminish the relevance and impact of a value assessment." The SAB therefore "advises the [Environmental Protection] Agency to identify the services and components of likely importance to the public at an early stage of a valuation and then to focus on characterizing, measuring, and assessing the value of the responses of those services and components to EPA's actions." The report further highlights the importance of a road map to guide valuation and recommends that each valuation should begin by "developing a conceptual model of the relevant ecosystem and the ecosystem services that it generates.

The goal of NESCS is primarily to support the first step in the process of CBA—that is, *identification* of policy-induced ecosystem service changes. Specifically, NESCS can be used to identify and categorize potential pathways through which policy-induced changes[7] to ecosystems ultimately result in human welfare changes. It provides a foundation that policy analysts can then use to conduct quantifications and valuations of ecosystem service changes in a consistent manner. In Section 1.2, we provide a brief review of basic terms and concepts. Section 1.3 describes the general approach for NESCS, and Section 1.4 summarizes some of the key requirements and a few key features of the system. We conclude Section 1 with a brief outline of the report in Section 1.5.

---

[6] It is important to note that green accounting involves evaluating the total value as opposed to changes to the system.

[7] Although the primary motivation for developing NESCS is to support evaluations of policies that cause changes to ecosystems, the framework can also be potentially useful for analyzing changes to ecosystems caused due to other factors such as natural changes that may occur over time.

1.2   Review of Basic Concepts

Since the focus of this report is to design and develop a classification system for ecosystem services, we begin by reviewing four important concepts. First, what is meant by a classification system and what are general principles or desirable characteristics of classification systems? Second, what does the term ecosystem service mean? Third, how are "services" generally defined for economic systems—how are services distinct from goods? Fourth, how are economic services distinct from ecosystem services? An understanding of all four concepts has important implications for the design of NESCS.

1.2.1   Classification Systems

The literature on taxonomies yields different definitions of and purposes for classification systems. Although the language varies, a common theme is that the primary purpose for a classification system is to provide an *organized structure*, through categories that allow one to group similar elements together and to separate different elements. Predetermined criteria define what should be considered similar or different, and these criteria are driven by the specific purpose for developing the classification system. One frequently cited definition is that a classification system is "the ordering or arrangement of objects into groups or sets on the basis of their relationships. These relationships can be based upon observable or inferred properties" (Sokal, 1974). The United Nations (UN) Department of Economic And Social Affairs (1999) defines general principles and best practices of classification systems, including:

- categories should be exhaustive and mutually exclusive;

- categories should be comparable to other international standard classifications;

- categories should be stable, meaning that they are not changed too frequently;

- the classification system should be well described and backed up by explanatory notes, coding indexes, coders, and other descriptors; and

- the classification system should be well balanced, that is., there should not be too many or too few categories.

1.2.2   Ecosystem Services

As mentioned earlier, a large variety of ecosystem service definitions and classification approaches have been proposed. These include de Groot et al. (2002), Millennium Ecosystem Assessment (MA, 2005), Wallace (2007), Boyd and Banzhaf (2007), Fisher and Turner (2008), Haines-Young and Potschin (2010a, 2010b, 2013), Staub et al. (2011), and Landers and Nahlik

(2013). Across these studies, there are differences in policy objectives, specific definitions of ecosystems services, and criteria for grouping services. There is general agreement that human well-being is supported by the existence, processes, and outputs of ecosystems, and that ecosystem services arise from this role. However, there is disagreement on the exact definition of ecosystem services. Specifically, studies disagree on the distinction between intermediate ecosystem processes/functions, final ecosystem services, and benefits; this lack of clear distinctions results in various issues and challenges for valuation. One assessment of the literature concludes that there is "a common lack of clarity in defining and valuing final ecosystem services, which has contributed to inconsistent valuations that double count some benefits and omits others" (Johnston and Russell, 2011).

1.2.3 Services in the Market

Given the differences in the definition of services in the context of natural systems, we explored how well-established economic accounting and classification systems define "services." We found that even in economic systems, services are difficult to define. For example, the Economic Classification Policy Committee (ECPC, 1993e) provides examples of alternative definitions available in the literature and concludes that "[t]here does not exist an internationally-agreed official definition of services…." One of the definitions included in ECPC (1993e) and adopted by the U.S. Census Bureau[8] is: "A service is a change in the condition of a person or a good belonging to some economic entity, brought about as a result of activity of some other economic entity…." Another definition of services provided on the website for the National Archives[9] is as follows: "A service is the production of an essentially intangible benefit, either in its own right or as a significant element of a tangible product, which through some form of exchange, satisfies an identified need. Sometimes services are difficult to identify because they are closely associated with a good; such as the combination of a diagnosis with the administration of a medicine." The website also notes that while goods (or "products") are "something that can be measured and counted, a service is less concrete and is the result of the application of skills and expertise towards an identified need."

We concluded, from our review of definitions of services in the economic context, that there is a general understanding and agreement that there are certain features of services that distinguish them from "goods." Unlike goods, **services** are typically intangible, non-storable, and inseparable from provider and consumer. Also, typically in economics, in contrast to goods,

---

[8]   http://www.census.gov/epcd/products/products99.htm (accessed May 29, 2015)
[9]   http://www.archives.gov/preservation/products/definitions/products-services.html (accessed May 29, 2015)

which can be treated as "stocks" and measured at a specific point in time, services[10] are viewed as "flows" from the provider to the consumer, and are measured over a period of time.[11]

1.2.4 Economic Versus Ecosystem Services

Since the focus of this report is on services provided by ecosystems/natural systems, it is important to compare and contrast them with services produced within economic/human systems. Some of the main differences are the following:

1. Market vs. Non-market nature of services: In contrast to economic services, ecosystem services are generally non-market in nature. In other words, they are typically not sold in markets and thus there are fewer observable transactions or prices.

2. Private vs. Public characteristics: Unlike economic services, ecosystem services often (although not always[12]) have "non-rival" characteristics; that is, enjoyment by one user does not diminish simultaneous enjoyment by other users.

3. Different implications of the concept of "final" services: Final *economic* services are sold to the end user—they flow from producers to households—whereas flows of final *ecosystem* services occur at the "point of direct hand-off" between natural systems and human systems (including both intermediate and final producers of economic goods, and households).

1.3   General Approach for NESCS

In designing the NESCS system, we have adapted concepts, principles, and methods from the different streams of literature described in Section 1.2. First, we attempt to incorporate the general principles and best practices of classification systems. Second, we draw from widely accepted concepts for classification and accounting of flows of services in the economic context. One of the key lessons learned from this literature is that services are defined as a flow rather than a stock. Third, we draw from previous literature on classification approaches for ecosystem services to address the question of where ecosystem services lie along the continuum and avoid double counting their values. Boyd and Banzhaf (2007) introduce and focus on the concept of

---

[10] Note that flows of ecosystem goods are not included or defined in the NESCS framework. For a detailed explanation, see Section 4.2.2.
[11] Goods can also be measured as flows, for example as the number of items produced in a year.
[12] Examples of exceptions would be water being drawn for drinking purposes since the water drawn cannot be used by others. We define categories for "extractive uses" (see Section 4.3.2) to account for these types of services.

"final" ecosystem services. As they define them, final ecosystem services occur at the point of hand-off between natural systems (ecosystems) and human systems (producers and households).

Realizing the importance of distinguishing between intermediate and final services and between stocks and flows, NESCS was designed to specifically focus on and classify **flows of final ecosystem services (FFES)**.[13] NESCS defines FFES as the direct contributions made by nature to human production processes or to human well-being.

Since services are viewed as flows from a provider to a consumer, in order to identify and define FFES, we first need to identify producers (or "supply-side") and consumers (or "demand-side") of the service. The two existing classification systems for economic goods and services in the United States (North American Industry Classification System, NAICS, and North American Product Classification System, NAPCS) also distinguish between supply-side and demand-side systems. [14] The NAICS system is designed to classify the production processes for goods and services based on a supply-side perspective (i.e., who is producing the commodities and how?), whereas the NAPCS system focuses on the demand-side perspective to classify the goods and services (i.e., how and by whom are the products being used?). We also make a distinction between a supply-side grouping and a demand-side grouping and thus include two complementary components, NESCS-S and NESCS-D when classifying FFES.

It is important to note is that while there are important parallels between NAICS/NAPCS and NESCS, there are important differences as well. Specifically, while NAICS and NAPCS provide alternative ways for classifying economic goods and services, NESCS-S and NESCS-D together constitute the classification system for FFES. They are complementary systems that need to be used together to identify and classify FFES. NESCS extends the NAICS/NAPCS framework to trace the flow of ecosystem services from natural systems to human systems.

The NESCS structure consists of four groups: (1) environmental classes that together cover the earth's surface; (2) classes of ecological end-products, which are the biophysical components of nature directly used or appreciated by humans; (3) classes of direct human use or non-use appreciation of end-products; and (4) classes of direct human users of end-products. The

---

[13] It is important to note that *flows of final ecosystem goods* are not included or defined in this framework. The main reason for this exclusion is that the process of transferring physical ecosystem products from nature to humans, which is necessary to generate flows of goods, typically requires **human** inputs (See Section 4.2.2 for more details).

[14] Both of these classification approaches were primarily designed to support the development of National Income and Product Accounts (NIPA). The NIPA are used to (1) trace the flow of *intermediate* goods and services between production sectors in the economy, and (2) estimate the value and composition of *final* goods and services sold to consumers.

first two groups pertain to the natural systems that "produce" FFES and can be interpreted as the supply-side classification (NESCS-S). The last two groups pertain to the human systems that appreciate or directly use FFES, and can be interpreted as the demand-side classification (NESCS-D). Combinations across these four groups define FFES, and can depict unique pathways that link changes in ecosystems with human welfare.

Within each of these four groups, NESCS adopts a nested hierarchical structure so that each group can be represented at multiple levels of aggregation or detail. NESCS provides a coding system that allows for a numeric representation of the NESCS structure. The categories in each of the four groups are assigned numeric codes. Each unique FFES can be referenced and identified by a NESCS code that can potentially be up to 15 digits.

1.4 Summary of Requirements and Key Features of NESCS

In this section we briefly summarize some of the basic requirements for marginal analysis and the key unique features of NESCS that will allow us to achieve our objectives. In order to support marginal analysis, it is important to have a standardized, comprehensive system that will allow for systematic linkages to be drawn between natural and human systems. It is important to ensure that there are no "leakages." In other words, the classification system should be such that there are no impacts of changes in policy that remain unaccounted for. At the same time, it is important to avoid double counting impacts. The following two complementary tools provided by NESCS help satisfy these requirements and help uniquely identify FFES:

- The first tool is the NESCS structure that defines categories and numeric codes for each of the four groups. These categories and codes are designed to help identify flows from ecosystems to human beings in a mutually exclusive way. Specifically, we define the supply-side and demand-side categories that can help provide linkages to ecological and valuation models respectively.

- The second tool is the NESCS conceptual framework that provides a way to systematically link and combine mutually exclusive categories from each of the four groups. It also provides a simplified framework for considering non-market (specifically environmental) sectors[15] (as represented by NESCS) and market sectors (as represented by NAICS/NAPCS[16]) in an integrated manner. This tool can also be used to represent

---

[15] Other sectors involving significant non-market elements include education and public sector services (NRC, 2005). These are not the focus of this report.

[16] Note that some inherently non-market activities are included in the NAICS/NAPCS sectors, such as owner-occupied housing and food consumed on farms (Nordhaus, 2004).

FFES pathways diagrammatically and in a readily understandable manner. It provides the linkages between different components of the framework, as between ecological production systems and market or non-market consumers.

Although it is not our main objective in designing NESCS, we also expect that NESCS can help to develop and support accounting systems such as green GDP. Therefore, we also provide a brief overview of the ways in which NESCS can aid these types of accounting systems:

- The NESCS conceptual framework provides a tool that can help differentiate between "intermediate" and "final" services, to avoid double counting. It can also help trace the input-output relationships between different sectors.

- The NESCS can help support green accounting in the following ways:

  - It strives to provide mutually exclusive and exhaustive categories to help avoid double counting.

  - It defines categories that can be used to present accounting data according to well-defined criteria.

  - It can help trace both sectoral and temporal changes, since it is based on a consistent and well-defined framework.

  - It may help in presenting accounts at different levels of aggregation due to its hierarchical structure.

  - It may help in adding services to accounts at a later time due to its flexible structure.

Before describing the details of the NESCS framework, classification structure, and coding system, it is important to draw the reader's attention to a few additional issues and features of the system. First, it must be emphasized that NESCS does NOT attempt to conduct quantification or valuation—the goal is to support identification of pathways between ecological and human systems.

Second, NESCS defines flows *from* natural systems *to* human systems, and not feedback effects from human to natural systems.[17] It is important to note that this is by design, and does not limit considering these dynamic and feedback effects when quantifying and valuing ecological benefits. However, considering these feedbacks is not essential for *defining and classifying* flows of ecosystem services from nature to humans. Although consideration of feedback and dynamic effects can be critical for policy analysis and valuation, they alter how NESCS is used but not how it is structured. Feedbacks may generate more flows through the NESCS system, which may require that more of the *existing* pathways in the system be considered. However, this does not imply that *new* pathways will need to be defined and classified to accommodate feedback effects.

Third, the NESCS framework describes and separates natural and human systems, but there are many "gray" areas. For example, in managed ecosystems like planted forests and national parks, determining "final" services is more challenging and requires more careful thinking since the natural and human systems overlap. It is, however more straightforward to define what is *not* an FFES. *In NESCS, anything that is produced using human inputs and sold in a market[18] is not considered an FFES.*[19]

Fourth, an important issue in classifying ecosystem services is the relationship between ecosystem services and human health. To be comprehensive, a classification system must cover all of the ways in which ecosystems contribute to human well-being; therefore, it must incorporate impacts on human health and safety. Ecosystems are vital for sustaining human life; however, the linkages between the environment and human health are both numerous and complex, including a wide range of direct and indirect pathways. To be comprehensive, NESCS is designed to account for these connections. However, rather than defining a separate ecosystem service category that exclusively addresses health effects, it defines multiple pathways that include human health or safety as key components. These pathways include, for example, direct uses and contact with air and water resources, protection against natural hazards, and indirect benefits from consuming health-enhancing goods and services (e.g., food, medicine, shelter) produced with ecological inputs.

---

[17] Only natural systems are capable of generating ecosystem services. Human intervention of any kind may change the profile of services that exist in any place, but the flow of ecosystem services originate through natural processes, or they would not meet standard definitions of ecosystem services.

[18] Not including regulatory-based environmental (i.e., cap-and-trade) markets.

[19] For example, agricultural landscapes are produced using human inputs and humans may have aesthetic appreciation for such landscapes. However, these landscapes are not sold in the market and may be considered to be externalities that result from agricultural production systems.

1.5   Overview of the Report

In the remainder of the report, we provide background on the relevant literature, describe the development and approach of the NESCS system, and provide a few illustrative applications. In Section 2, we provide a review of the literature on classification approaches to ecosystem services. We do not attempt to be exhaustive in this review. Rather, we focus on a few key studies to help provide understanding of some of the main concepts and issues that are relevant for our design. Section 3 provides a brief overview of economic accounting and classification systems. This provides important background for our approach, since NESCS draws from the principles and structure of these systems. In Section 4, we describe the NESCS methodology. Specifically, we describe our approach in detail, develop the conceptual framework, and describe the NESCS structure and coding system in detail. Illustrative policy examples are used in Section 5 to demonstrate how NESCS may be applied in practice. Section 6 concludes with a summary of key design elements and features. It also provides a short comparison with other classification systems, primarily Landers and Nahlik (2013). It then identifies other potential applications and next steps for future research.

SECTION 2
REVIEW OF ECOSYSTEM SERVICES CLASSIFICATION LITERATURE

2.1   Introduction

Since the publication of *Nature's Services* by Gretchen Daily (1997), a growing body of literature has emerged on classifying ecosystem services. In this section we summarize the recent research in this area. All of the studies we review share the same fundamental purpose, which is to identify and describe the various ways in which ecosystems support human welfare. However, they also provide different perspectives, using different approaches and terminology to address this common purpose. The studies selected for review in this section include papers and reports from both peer-reviewed and gray literature. They specifically include the following studies: Daily et al. (1997), de Groot et al. (2002), Millennium Ecosystem Assessment (MA) (2005), Wallace (2007), Boyd and Banzhaf (2007), Fisher and Turner (2008), Haines-Young and Potschin (2010a, 2010b, 2013), Staub et al. (2011), and Landers and Nahlik (2013).[20]

The selected studies differ in three main respects. First, the structure and level of detail of the classification systems vary across studies and have generally evolved over time. They range from a "flat" structure, which mainly provides a list of ecosystem services, to more complex hierarchies and taxonomies, which provide multi-level embedded groupings of ecosystem services. Two of the studies—Boyd and Banzhaf (2007) and Fisher and Turner (2008)—define principles for classifying ecosystem services rather than providing an explicit classification system. The initially proposed hierarchies use "functional groupings" (i.e., they grouped similar ecosystem functions under the same category) (de Groot et al., 2002; MA, 2005), whereas later studies suggest and implement groupings based on benefits to humans (Wallace, 2007; Staub et al., 2011). Another development in the more recent literature is to use a flexible nested hierarchy that allows for easy aggregation at different levels and for incorporation of additional services when they become relevant (Haines-Young and Potschin, 2010a, 2010b, 2013; Landers and Nahlik, 2013).

Second, the studies address different policy analysis objectives. These differing objectives account for many of the distinct features adopted across classification systems. The primary objective in the initial studies was to list and describe the ways in which ecosystems support human welfare (Daily et al., 1997; de Groot et al., 2002) and to expand on this list to make it exhaustive (MA, 2005). The analytical objectives in the more recent studies include

---

[20] Examples of other studies include Costanza et al. (1997), National Research Council (2005), Turner et al. (1994), Hawkins (2003), and Hein et al. (2006). The last two approaches categorize ecosystem services into those that have use values and those that have non-use values.

support for natural resource management decision making, green accounting, and cost-benefit analysis. For natural resource management decisions, a well-defined and internally consistent classification system is essential for making meaningful comparisons between different courses of action (Wallace, 2007). For example, it is difficult to make a well-informed choice between alternative water quality protection approaches if one of them is evaluated in terms of sediment retention (a process) and the other in terms of changes in water clarity (the outcome of a process).

From an ecosystem services perspective, one of the key differences between cost-benefit analysis and green accounting is that cost-benefit analysis typically requires value estimates for changes in ecosystem services due to a policy, program, or management action (Fisher and Turner, 2008), whereas green accounting focuses more on the level of ecosystem service flows provided by natural systems during a given time period. For example, the objective of green GDP accounting is to provide an annual monetary measure indicative of national welfare that includes contributions from natural systems.

Third, and perhaps most importantly, the studies differ in how they define the concept of ecosystem services. Generally speaking, the different classification systems are consistent with the conceptual framework shown in **Figure 2-1**. In particular, there is general agreement that ecosystems provide benefits to humans in various ways, including (1) through the outputs of their own processes and functions (e.g., climate regulation), and (2) by supporting human activities (e.g., food production) that then provide benefits. A main area of disagreement, however, is where "ecosystem services" occur along the continuum between ecosystems and human welfare. In particular, the classification approaches differ in whether natural processes or functions should themselves be considered services and whether services and benefits should be treated as synonymous. They also differ in whether ecosystem services should include items that involve input from humans (e.g., food production that requires human labor inputs) or whether these services must inherently be delivered from natural processes or components prior to human involvement (e.g., unmanaged pollination). Therefore, even though a consensus has emerged in the literature about the importance of differentiating "final" ecosystem services from the "intermediate" processes that contribute to them, there is less agreement about what constitutes a final service.

Figure 2-1. Simple Conceptual Framework Underlying Most Ecosystem Service Definitions and Classification Systems

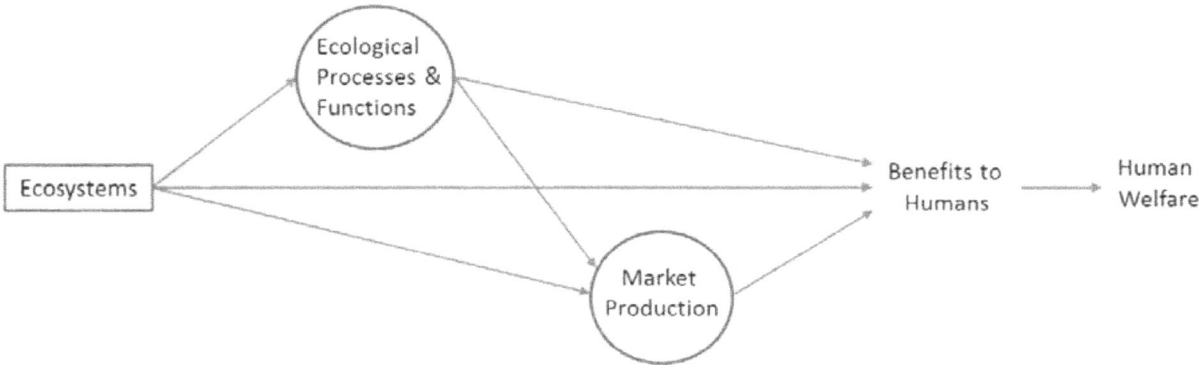

Despite these differences, the literature has converged on several key points, including the following:

1. There is a general recognition that humans benefit from ecosystem services through both market and non-market activities.

2. For accounting or valuation purposes, it is important not to count both intermediate and final services (or intermediate process and final outcome), to avoid double counting values.

3. Overlapping categories of ecosystem services must be treated with caution to avoid or minimize double counting.

4. What may be an intermediate service for one category of benefit may be a final service for a different category.

5. Values for ecosystem services may be location specific, time specific, and consumer specific.

Given these areas of agreement, the evolution in the literature has involved attempts to rigorously define ecosystem services in a way that:

- is consistent and meaningful across different type of services;

- is measurable and operational;

- helps to avoid double counting services; and

- corresponds to the context and objectives of the study.

13

The earlier studies (e.g., Daily et al., 1997; MA, 2005) were mainly devoted to developing an inventory of ecosystem services without specific attention to avoiding overlaps.[21] The later studies (e.g., Wallace, 2007; Boyd and Banzhaf, 2007; Fisher and Turner, 2008) critique different aspects of the earlier definitions and suggest alternative ways to address them to meet the four main goals above. In particular, Haines-Young and Potschin (2010a, 2010b, 2013) and Landers and Nahlik (2013) define specific pathways through which ecosystems provide services to human beings, and Staub et al. (2011) provide a way to implement the definitions with data relevant to the Swiss economy.

In the following sections, we describe how the ecosystem services term has been defined in each of the studies. We then describe the associated classification system (or some key examples if a complete classification is not presented), as well as the main advantages, shortcomings, and criticisms of each approach.

2.2 Daily et al. (1997)

2.2.1 Objective

One of the seminal papers in the ecosystem services literature, the main objective of Daily et al. (1997) was to identify and describe the main connections between ecosystems and human well-being.[22]

2.2.2 Definition and Discussion

The paper defines ecosystem services as "a wide range of conditions and processes through which natural ecosystems, and the species that are a part of them, help sustain and fulfill human life. They maintain biodiversity and the production of ecosystem goods, such as seafood, forage timber, biomass fuels, natural fiber, and many pharmaceuticals, industrial products, and their precursors" (p. 2). Some examples of such services are:

- purification of air and water;

- mitigation of droughts and floods;

- generation and preservation of soils and renewal of their fertility;

- detoxification and decomposition of wastes;

---

[21] MA (2005) recognizes that some of their categories overlap (see Section 2.4).

[22] Additional and more detailed discussions of the issues raised in this paper are provided in the book *Nature's Services* (Daily, 1997)

- pollination of crops and natural vegetation;

- dispersal of seeds;

- cycling and movement of nutrients;

- control of the vast majority of potential agricultural pests;

- maintenance of biodiversity;

- protection of coastal shores from erosion by waves;

- protection from the sun's harmful ultraviolet rays;

- partial stabilization of climate;

- moderation of weather extremes and their impacts; and

- provision of aesthetic beauty and intellectual stimulation that lift the human spirit.

This approach highlights the fact that natural ecosystems provide market goods as well as non-market services.

Daily et al. (1997) also emphasize the importance of the spatial nature of ecosystem services. They point out that the "[f]low of ecosystem goods and services in a region is determined by type, spatial layout, extent and proximity of ecosystems supplying them" (p. 6). This spatial dimension has important implications for valuation of ecosystem services. For example, the value of the flood prevention services offered by a wetland depends critically on its location within a floodplain in relation to vulnerable populations and ecosystems.

2.2.3 Limitations

While recognizing that the objective of Daily et al. was to link ecology and human well-being (as opposed to generating an accounting system), Boyd and Banzhaf (2007) point out the problems that can arise when using this definition of ecosystem services in an accounting framework. These problems occur because some of the described services may be better characterized as ecosystem processes or functions than as final ecosystem services. For example, even though water purification is embodied in the production of clean water, the service itself is clean water. As a result, double counting ecosystem services may occur if no distinction is made between the intermediate processes and the final service. In addition, if the aim is to provide a measurement of ecosystem services, problems can arise in applying this framework because the measurement of processes is typically more difficult than the measurement of outcomes of processes. Boyd and Banzhaf also draw a distinction between "benefits" and "services" provided by ecosystems, and argue that some of the items defined by Daily et al. as ecosystem services are in fact benefits. For example, they argue that flood control is a benefit to which natural assets (e.g., wetlands) contribute, not a service.

2.3   de Groot et al. (2002)

2.3.1   Objective

The 2002 study by de Groot et al. notes that although a substantial amount of research has been done on the value of ecosystem services, the resulting data are not necessarily defined at compatible scales of analysis and are classified differently. The goal of this de Groot et al. study is to support comparative ecological economic analyses. To support this, the authors present a "conceptual framework and typology for describing, classifying and valuing ecosystem functions, goods and services" (p. 393).

2.3.2   Definition and Discussion

The authors emphasize the importance of translating complex ecological structures and processes to a limited number of ecosystem functions. They define ecosystem functions as "the capacity of natural processes and components to provide goods and services that satisfy human needs, directly or indirectly." Ecosystem functions thus are antecedents to ecosystem goods and services.

The paper groups 23 ecosystem functions and their associated ecosystem goods and services into four broad categories:

1. Regulation Functions—includes the biogeochemical cycles (nutrient, carbon, water, etc.), that support and maintain life. Ecosystem services derived from this function category also include clean air, water, soil, and disturbance prevention. Eleven separate ecosystem functions are described as falling under regulation functions.

2. Habitat Functions—habitat for wild plants and animals is necessary for the services the plants and animals provide through production and information functions described below. Two distinct ecosystem functions are described—refugium and nursery function—to illustrate the importance of different habitat requirements across life histories.

3. Production Functions—food, medicine, and materials produced by ecosystems. Excluded from this category are nonrenewable resources, such as gold or oil. Also included in this category are genetic resources that can replenish domesticated species.

4. Information Functions—recreation, aesthetic, cultural, spiritual, scientific, and educational services.

By their definition, while regulation functions provide some ecosystem services, most are generated through the production and information functions, with regulation and habitat functions providing the necessary inputs for production and information functions.

To link ecosystem functions, goods, and services to values, the authors describe three separate categories of value. Ecological value is included in their framework to place a "sustainable use" limit on the consumption of ecosystem goods and services. Socio-cultural value corresponds well to equity considerations, such as environmental justice, when assessing ecosystem service values. Finally, the category of economic value is where actual money values may be assigned to the ecosystem goods and services provided by ecosystem functions.

2.3.3 Limitations

Wallace (2007) criticizes de Groot et al. for combining "processes (means) for achieving services and the services themselves (ends) within the same classification category" (p. 236), and for the risks of double counting benefits. Similarly, Wainger and Mazzotta (2011) note that valuing functions and processes may lead to double counting. They also recommend exclusion of basic ecological functions and processes such as nutrient cycling, for which people do not have well-established preferences.

The study by de Groot et al. has also been criticized for their use of the term "ecosystem functions" as the subset of ecosystem processes that provide ecosystem services. Wallace (2007) finds the use of the term to be redundant to ecosystem process, and advocates not using ecosystem function for greater parsimony of terms and for clarity. Similarly, Haines-Young and Potschin (2010a, 2010b, 2013) recommend dropping the term ecosystem function, which is often used more generally than de Groot et al.'s definition, and instead using the term "capability."

2.4   Millennium Ecosystem Assessment (2005)

2.4.1   Objective

The goal of the Millennium Ecosystem Assessment report (MA, 2005) was to "establish the scientific basis for actions needed to enhance the contribution of ecosystems to human well-being without undermining their long-term productivity" (p. ii) Thus the objective of their conceptual framework is to assess the consequences of changes in ecosystems for human well-being.

2.4.2   Definition and Discussion

The MA report defines ecosystem services in the following way: "Ecosystem services are the benefits people obtain from ecosystems. These include provisioning services such as food and water; regulating services such as flood and disease control; cultural services such as spiritual, recreational, and cultural benefits; and supporting services, such as nutrient cycling, that maintain the conditions for life on Earth" (p. 39).

The MA report also refers to other types of categorizations of ecosystem services that have been proposed in the literature:

- functional groupings, such as regulation, carrier, habitat, production, and information services (de Groot et al., 2002);

- organizational groupings, such as services that are associated with certain species, that regulate some exogenous input, or that are related to the organization of biotic entities (Norberg, 1999); and

- descriptive groupings, such as renewable resource goods, nonrenewable resource goods, physical structure services, biotic services, biogeochemical services, information services, and social and cultural services (Moberg and Folke, 1999).

The MA report opts for a functional groupings approach and uses categories of provisioning, regulating, cultural, and supporting services. The first three categories directly affect people and supporting services, and the supporting services are needed to maintain the other services. These categories, along with the impact they have on human well-being, are displayed in Figure 2-2. The figure depicts supporting services that differ from provisioning, regulating, and cultural services. Their impacts on people are either indirect or occur over a long time, whereas changes in the other categories have relatively direct and short-term impacts on people. Thus, supporting services do not contribute directly to welfare. For example, although humans do not directly use soil formation and retention services, changes in these services do indirectly affect people through their impact on the provisioning service of food production. Some other examples of supporting services are primary production, production of atmospheric oxygen, nutrient cycling, water cycling, and provisioning of habitat.

Figure 2-2. MA Categorization of Ecosystem Services and their Links to Human Well-Being

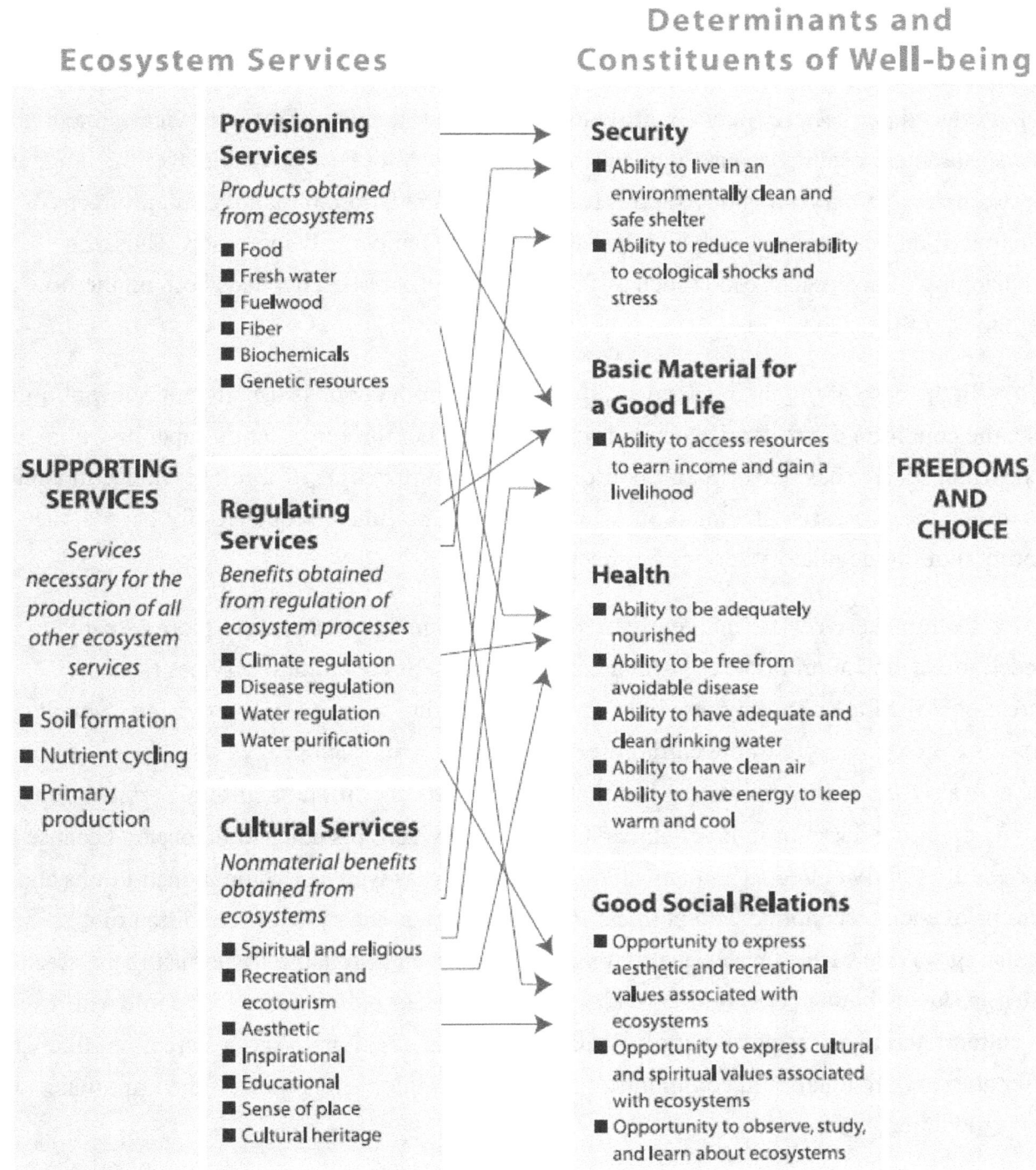

The MA report recognizes that there are overlaps between these categories of ecosystem services; however, its developers argue that "the purpose is not to establish a taxonomy but rather to ensure that the analysis addresses the entire range of services" (MA, 2003, p. 38). For example, erosion control can be categorized as both a supporting and a regulating service, "depending on the time scale and immediacy of their impact on people" (p. 40).

The MA report also describes what they would consider to be reasonable indicators or measures of the condition of services for three of the defined categories. For provisioning services, flow measures alone (e.g., biophysical production measured in terms of kilograms of crop produced per hectare) may not provide an accurate reflection of these services. Potential mismeasurement occurs because a given flow may or may not be sustainable over the long term. For example, overharvesting of fisheries leads to the degradation of the long run productive potential of the resource, even though harvest may have temporarily increased. Thus, the provisioning of ecological goods such as food, fuel wood, or fiber, depends both on the flow and the "stock" of the good.

In the case of regulating services, the level of "production" is usually not relevant. In this case, the condition of the service depends more on whether the ecosystem's capability to regulate a particular service has been enhanced or diminished. For example, if forest clearance in a region has resulted in decreased precipitation, which has had harmful consequences for people, then the condition of that regulatory service has been degraded.

Cultural services are inherently more difficult to measure; therefore more research is needed to develop appropriate measurement approaches. Some cultural services (such as recreational fishing or hunting) are linked to a provisioning service (food provision), which can serve as a proxy measure of the cultural service. However, no such proxy exists for most cases. The MA also draws attention to the fact that assessing the condition of cultural services depends heavily on either direct or indirect human use of the service. This connection occurs because the services are tightly bound to human values and behavior, as well as to human institutions and patterns of social, economic, and political organization. For example, the condition of a regulating service such as water quality might be high even if humans are not using the clean water produced, but an ecosystem provides cultural services only if there are people who value the cultural heritage associated with it. Thus, perceptions of cultural services are more likely to differ among individuals and communities than, for example, perceptions of the importance of food production.

For supporting services, the MA report concludes that a normative scale for assessing the condition of these services is not always practical, since the link to human benefits is indirect.

2.4.3 *Limitations*

Boyd and Banzhaf (2007) argue that although the MA report does attempt to motivate measurement (and is successful for provisioning services), problems arise in the case of the non-marketed components. Similar to the Daily et al. classification, many of the regulating services correspond to better to functions and processes than to services (e.g., pest regulation, disease regulation, hazard reduction, pollination, and climate regulation). Also, they argue that cultural services, including spiritual and religious values, aesthetic values, and recreation and ecotourism, should be characterized as benefits rather than as services.

Haines-Young and Potschin (2010a, 2010b, 2013) also note that there is a distinction between benefits and services for two reasons. First, most services have multiple benefits. For example, food provides health, pleasure, and sometimes even cultural identity. Second, similar to criticism by Boyd and Banzhaf (2007), they note that benefits have a human component to them. Wallace (2007) offers a similar criticism of the MA framework and other classification systems for mixing ecosystem processes (means) with ecosystems services (ends). According to this critique, most of the services under the regulating and supporting categories are processes rather than services. This critique also implies that regulating and supporting services are not at the same level[23] as, say provisioning services, and therefore cannot be compared easily with each other and traded off in a decision system.

2.5 Boyd and Banzhaf (2007)

2.5.1 *Objective*

Boyd and Banzhaf (2007) highlight the importance of standardized ecosystem units in the development of national-scale environmental accounting systems. Their objective is to develop an ecosystem services framework that is "potentially consistent with national income accounting and hence a broad 'green GDP'" (p. 617).[24]

---

[23] The meaning of this term will be explained further in a later section.

[24] Boyd and Banzhaf also indicate that along with definition of quantities, accounting frameworks require aggregation and weighting. They suggest that estimates of willingness to pay (WTP) from non-market valuation studies can be used for weighting. To ensure that WTP–based weights are spatially explicit, meta-analysis of existing values can be used to calibrate benefit transfers. This would involve using WTP indicators (where WTP are functions of landscape indicators) along with site-specific GIS measures of ecosystem scarcity, substitutes, and complements. However, they note that such information is not readily available.

2.5.2 *Definition and Discussion*

Boyd and Banzhaf's definition is as follows: "Final ecosystem services are components of nature, directly enjoyed, consumed, or used to yield human well-being" (p. 619). One important implication of this definition is that ecosystem services are not benefits, nor are they always the final product consumed. For example, it is more appropriate to consider recreation as a benefit rather than an ecosystem service, since it is produced using both ecological services and conventional goods and services. Recreational angling (or fish caught) is a function of ecosystem services like fish population, as well as human inputs such as travel and fishing equipment.

This definition makes a distinction between intermediate and final products to avoid the problem of double counting. For example, different ecosystem components contribute to clean drinking water (which is consumed directly by households), but according to this framework, it is the clean water, rather than the ecosystem structures or processes producing the clean water, that should be included in an ecosystem service account. Also, an ecosystem component may be a final service in one context and an intermediate service in another. For example, whereas clean water is a final ecosystem service for drinking water benefits, it is an intermediate component for sustaining fish populations, which are one of the final ecosystem services needed to support recreational fishing benefits.

According to this framework, services are ecological things or characteristics, not functions or processes. "Ecosystem processes and functions are the biological, chemical, and physical interactions between ecosystem components. Functions and processes are not end-products; they are intermediate to the production of final ecosystem services" (p. 620). For example, nutrient cycling is an ecological function, not a final service.[25] Pollination is a process, while delivery of pollen is the service. One of their observations is that narrowing the range of things to be counted (by monitoring the end-products rather than complex ecological processes) helps establish priorities for limited data-collection budgets.

Similar to national accounts, which are more of a proxy for components of welfare rather than for welfare itself, Boyd and Banzhaf state that an ecosystem measure should be thought of as "a measure of nature's value, not the value itself" (p. 617). Thus, this framework views ecosystem services as a welfare indicator. They point out that in a welfare accounting framework like green GDP, it is critical to distinguish between prices and quantities. To consistently measure changes in welfare over time, one must hold prices of goods and services fixed and

---

[25] An analogous example for marketed goods and services would be that of a manufacturing process. The value of this is not included in GDP, as its value is embodied in the value of its end-products.

measure changes in the quantities of these goods and services. Applying this principle to green GDP accounting requires that ecosystem services be measured as quantities. In contrast, they argue that in a cost-benefit framework the separation of prices and quantities is less important, since it is the measure of "total benefits" (i.e., the product of price and quantity) that matters.

Although Boyd and Banzhaf do not attempt to provide a complete inventory of services, they outline a procedure for developing such an inventory. The first step would be to inventory sources of well-being related to nature (e.g., aesthetic enjoyment, various forms of recreation, maintenance of human health, physical damage avoidance, and subsistence or foraged consumption of food and fiber). Once these sources are identified, final ecosystem services can be identified as the ecological end-products that can be used to produce the well-being. For example, natural land cover in a viewshed is an end-product that contributes to aesthetic enjoyment and outdoor recreation, and fish populations are an end-product that contributes to subsistence food consumption.

2.5.3 Limitations

Fisher and Turner (2008) disagree with one of the key aspects of the above definition, which is that services are viewed as ecological components, that is, countable things such as lakes, forests, or fish populations. According to Fisher and Turner (2008), functions and/or processes are ecosystem services as long as there are human beneficiaries. They state that "This is important because it connects human welfare to nature throughout an ecosystem, not just the endpoint" (p. 1168). Like Costanza (2008a), Fisher and Turner (2008) disagree with the paper's assertion that only direct endpoints can be a service. According to their different interpretation, "as long as human welfare is affected by ecological processes or functions (somewhere down the line) they are services" (p. 1168).

2.6 Wallace (2007)

2.6.1 Objective

The goal of Wallace (2007) is to provide an ecosystem service framework for natural resource management decisions. To support this objective, ecosystem services must be classified "in a way that allows comparisons and trade-offs among the relevant set of potential benefits" (p. 236).

2.6.2 Definition and Discussion

Wallace defines ecosystem services using the terminology from MA (2005)—as "the benefits people obtain from ecosystems" (p. v). These benefits include food, water, timber, cultural values, and others, and are the outcomes sought through ecosystem management.

The relationship between ecosystem processes and the structure and composition of natural elements in the ecosystem is best described as a sequential process. In the first period, the structure and composition of ecosystems are modified by ecosystem processes to create a new structure and composition in the second period. The structure and composition are measured at different points in time to derive the change in processes and to quantify the distribution of ecosystem elements at these points. The role of natural resource managers is to maintain or change the ecosystem elements to better support human values. This approach argues, however, that ecosystem services should be described in terms of the structure and composition of ecosystems rather than in terms of ecosystem processes.

One of the reasons provided for using structure and composition of ecosystems to describe ecosystem services is similar to Boyd and Banzhaf (2007)—the former are more easily observed and measured, more sensitive to degradation and less expensive to monitor. Also, some species may be irrelevant to key processes, but their extinction would indicate a loss of an ecological asset. Finally, human well-being, whether tangible or intangible is measured in quantities rather than whether, for example, carbon or nitrogen cycles are working adequately. Thus, the task of natural resource managers is to influence ecosystem processes to ensure that the composition and structure of ecosystem elements continuously delivers human well-being.

Another issue raised by Wallace (2007) is the importance of determining the point at which processes deliver ecosystem services. This determination helps to avoid double counting. Similar to Boyd and Banzhaf, he defines this as the point at which an ecosystem directly provides an asset used by one or more humans, which is the relevant end of a causal chain and provides the delivery of a service. Thus, this approach does not mix "means" and "ends." All ecosystem assets are defined at the same "level," in that they are all directly used or otherwise of benefit to individual humans.

Wallace then notes that "it is possible to examine trade-offs and other aspects of decisions at either the level of services or the level of values. This should not cause difficulties provided decisions are made among either services or values, not a mixture of both, and the set of endpoints chosen are relevant to the goal driving the decision" (p. 240). Again, similar to Boyd and Banzhaf, he points out that the "broad components that make up a reasonable quality of life are likely to be consistent across cultures, but the relative weighting, specification and means of achieving these components will vary among cultures and among individuals from any one culture" (p. 240).

Wallace also proposes an alternative classification approach, where services are grouped according to the human values they support. Thus, services are described in terms of the structure and composition of particular ecosystem elements (expressed as assets), and these services are in turn classified according to the specific human values they support. The categories of human values include: adequate resources; benign physical and chemical environment; protection from predators, disease and parasites; and sociocultural fulfillment.

2.6.3   Limitations

One of the criticisms offered by Fisher and Turner (2008) is that the Wallace framework considers services and benefits to be the same. For valuation, this lack of distinction is a problem and could lead to a problem of double counting. "For example, adding values for primary production to values for recreational hiking would 'double count' the value that say forests add to the hikers' experience" (p. 1168). The other problem is that benefits like recreation could include non-ecological (i.e., human) components, which are not appropriate to be considered as an ecosystem service.

Fisher and Turner (2008) and Costanza (2008a) disagree with Wallace's assertion that only direct endpoints can be a service. According to them, to avoid double counting, intermediate services should not be added to final services, but they should nonetheless be interpreted as services.

Costanza (2008a) also argues that Wallace is essentially trying to differentiate between final and intermediate, rather than between "means" and "ends" ("end" to him is welfare and "services" are a "means" to achieve that end). Wallace (2008) responds to Costanza by emphasizing that the distinction between "means" and "ends" has implications other than those between "final" and "intermediate" services.[26] He points out that in order for a manager to choose between two options, they would have to be readily comparable. For example, how would a natural resources manager trade off pollination with clean water (provisioning)? Costanza and Wallace also do not agree on the necessity of a single consistent framework for ecosystem services.

---

[26]   See Wallace (2008) for illustrative examples.

2.7   Fisher and Turner (2008)

2.7.1   *Objective*

Fisher and Turner (2008) is a response to Wallace (2007). It critiques Wallace's proposed framework, as well as the approaches in MA (2005) and Boyd and Banzhaf (2007). They point out that these approaches are based upon the context in which they are being used as well as the definition being used. They argue that although each approach is suitable for its own purpose, none of these approaches are suitable for determining: (1) how ecosystem services deliver human welfare benefits; (2) where the benefits are realized; (3) by whom the benefits are enjoyed; and (4) how their value changes across the landscape under different future scenarios. To address these limitations, they propose an alternative definition and classification approach for ecosystem services.[27]

2.7.2   *Definition and Discussion*

Fisher and Turner's definition of ecosystem services (which draws largely on Boyd and Banzhaf) is as follows: Ecosystem services are "the aspects of ecosystems utilized (actively or passively) to produce human well-being" (p. 1168). They point out three key features of their definition and compare them with the other three studies. Their conclusions are summarized in Table 2-1. The main distinction they draw with respect to the Boyd and Banzhaf approach has to do with whether ecosystem functions and processes (e.g., flood regulation, nutrient cycling) should be included in the definition of ecosystem services. Fisher and Turner argue that they should be included as intermediate ecosystem services.[28] An illustrative example showing a relationship between some intermediate services, final services, and benefits is provided in Table 2-2.

---

[27]   Additional details of this proposed approach are provided in Fisher, Turner, and Morling (2009).

[28]   Boyd and Banzhaf (2007) do not necessarily reject this idea, but they use the terms "intermediate ecological components" and "intermediate ecological processes" rather than "intermediate ecological services."

Table 2-1. **Characteristics of Fisher and Turner's (2008) Definition and Comparison with Other Classification Systems[a]**

	Characteristic of F&T 2008 Definition		Similarity/Dissimilarity with other Studies
1	Services are not benefits	Similarity with Boyd and Banzhaf	"As deftly pointed out by Boyd and Banzhaf (2007) services and benefits are different. They argue that recreation is not a service provided by ecosystems, but rather a benefit of which ecosystems provide important inputs. A benefit is something that has an explicit impact on changes in human welfare, like more food, better hiking, less flooding."
		Dissimilarity with Wallace and MA	"Wallace (2007) and the MA (2005) consider services and benefits to be the same. For valuation, this is a problem and could lead to a problem of double counting. For example, adding values for primary production to values for recreational hiking would ''double count'' the value that say forests add to the hikers experience."
2	Ecosystem services are ecological in nature	Similarity with Boyd and Banzhaf	"Again, similar to Boyd and Banzhaf in that aesthetic values, cultural contentment and recreation are not ecosystem services. They are benefits, and are not just a function of ecosystems, but include other inputs like human capital, built capital, etc. They are benefits also because they directly relate to changes in human welfare."
		Dissimilarity with Wallace and MA	"For Wallace (2007) and the MA (2005) these things [aesthetic values, cultural contentment and recreation] are services."
		Dissimilarity with Boyd and Banzhaf	"We differ here with Boyd and Banzhaf (2007) in that they see services as ecological components, i.e., things you can count like lakes, forests, fish populations. We think that functions and/or processes are ecosystem services as long as there are human beneficiaries. This is important because it connects human welfare to nature throughout an ecosystem, not just the endpoint."
		Similarity with Daily and MA	"This is in line with Daily (1997 and the MA (2005) which both make this connection explicit through the word service, not obscure it in ecological lexicon (i.e., processes, functions). For example, flood regulation is an ecosystem service here, as in Daily (1997) and the MA (2005), but is considered a process in Boyd and Banzhaf (2007) and Wallace (2007)."
3	Ecosystem services do not have to be utilized directly	Dissimilarity with Boyd and Banzhaf and Wallace	"Here we take the opposite view of Boyd and Banzhaf (2007) and Wallace (2007) who argue that only the direct endpoints are ecosystem services. We argue that as long as human welfare is affected by ecological processes or functions (somewhere down the line) they are services."
			"[P]ollination is an ecosystem service since it is an ecological phenomenon that we utilize (indirectly) to enjoy certain food benefits. For us it makes more sense to call pollination an ecosystem service than say the almonds that we benefit from. In both Boyd and Banzhaf (2007) and Wallace (2007) it would be almonds that are the ecosystem service."
		Similarity with Daily	"Carbon sequestration is an ecosystem service because there are net human benefits derived for this process in a world of changing climate. This is in line with much of Daily's original text (1997)."

[a] All quotes are from p. 1168 in Fisher and Turner (2008)

Table 2-2. Illustrative Example of Relationships Between Some Intermediate Services, Final Services, and Benefits (Fisher and Turner [2008])

Abiotic Inputs	Intermediate Services	Final Services	Benefits
Sunlight rainfall nutrients, etc.	Soil formation Primary productivity Nutrient cycling	Water regulation	Water for irrigation Drinking water Electricity from hydro-power
	Photosynthesis Pollination Pest regulation	Primary productivity	Food Timber Nontimber products

Fisher and Turner point out that there are multiple relationships between ecosystem processes and human benefits; however, there is little risk of double counting in valuation exercises, as only distinct benefits are valued. The key factor for ecosystem service research is that project scientists and stakeholders agree on the "line between final services and benefits, so that we can manage, monitor and make policy to protect services that help maintain (and/or value) that benefit" (p. 1169).

Fisher and Turner also note that their approach is consistent with other features of ecosystem services (described in above sections). For example, the delineation between intermediate services, final services, and benefits is not strict. Services are often a function of a beneficiary's perspective. Also, the same service can generate multiple benefits—for example, water regulation provides flood prevention, drinking water, and recreation potential. It would be appropriate to add the value of these benefits together since these are distinct benefits. Finally, characteristics like resilience and functional diversity would also be services under this framework, as they are ecological phenomena from which humans derive benefits. However, defining the benefit would require modeling and scenarios to understand just what the benefit from such a service is, so it would remain difficult to attach meaningful economic valuation.

2.7.3 Limitations

Although Fisher and Turner propose an alternative conceptual framework for defining and classifying ecosystem services, they do not apply this framework to develop or specify a formal alternative classification system for ecosystem services. Instead, they provide illustrative examples to explain how they differentiate between intermediate services, final services, and benefits. A more detailed and extensive application of their proposed framework is needed to fully evaluate its usefulness as a basis for ecosystem services classification.

2.8 Roy Haines-Young and Marion Potschin (2010a, 2010b, 2013): Common International Classification for Ecosystem Services (CICES)

2.8.1 Objective

The proposal for a Common International Classification for Ecosystem Services (CICES) is summarized in Haines-Young and Potschin (2010a, 2010b, 2013). One of the goals of this classification system is to be consistent with accepted typologies of ecosystem goods and services currently being used in the international literature, and to be compatible with the design of Integrated Environmental and Economic Accounting methods. CICES is motivated by some essential directives in the report of the EPA Science Advisory Board (EPA, 2009):

1. It is important to identify relevant ecosystem services as a common list that can serve different purposes.

2. Classification methodologies should follow some basic principles.

3. It is essential that classifications should help us avoid the problem of double counting and so provide the basis for accurate assessments and valuations.

4. The contributions ecosystems make to human well-being should be defined in terms that are both concrete and meaningful to those whose lives are affected by them.

2.8.2 Definition and Discussion

In the proposal for CICES, ecosystem goods and services are defined to be the contributions that ecosystems make to human well-being, and arise from the interaction of biotic and abiotic processes.[29]

To resolve the problem of identifying concrete outcomes, this approach seeks to cross-reference ecosystem services with existing classifications of products and services, so that the contributions that ecosystems make in the form of services can be better identified and quantified. To be able to link changes in ecosystem structures and processes to economic consequences, Haines-Young and Potschin (2010a, 2010b, 2013) note that it is also essential to link ecosystem services and land cover.

Figure 2-3, taken from Haines-Young and Potschin (2010a), depicts a "pathway" from ecosystems to human well-being. Haines-Young and Potschin (2010a) describes the key features

---

[29] Haines-Young and Potschin (2010a, 2010b, 2013) note the distinction between biodiversity and geodiversity and note that in the proposed CICES classification, both biotic and abiotic elements (including minerals, wind, snow, salt, etc.) would be included when defining a service.

of this "production chain," such as ecosystem components, structures, processes, functions, services, and benefits. They note that although some of these terms have been interchangeably used in the literature, it is important to make distinctions between certain concepts. However, they also emphasize that whatever terminology is used, a mix of structures, processes, and function generates the services that ultimately provide benefits to people. "Thus, services are best seen as the 'useful things' ecosystems 'do' for people in relation to enhancing human well-being directly or indirectly, and that we should strive to be clear about what we label as a service and how it is to be measured and valued" (p. 7). In light of the above, CICES seeks to identify only the "final products" of ecosystems and thus includes what was termed as provisioning, regulating, and cultural services in the MA (2005).

Figure 2-3. Defining Ecosystem Functions, Services, and Benefits, and the Context for CICES (Source: Haines-Young and Potschin, 2010a)

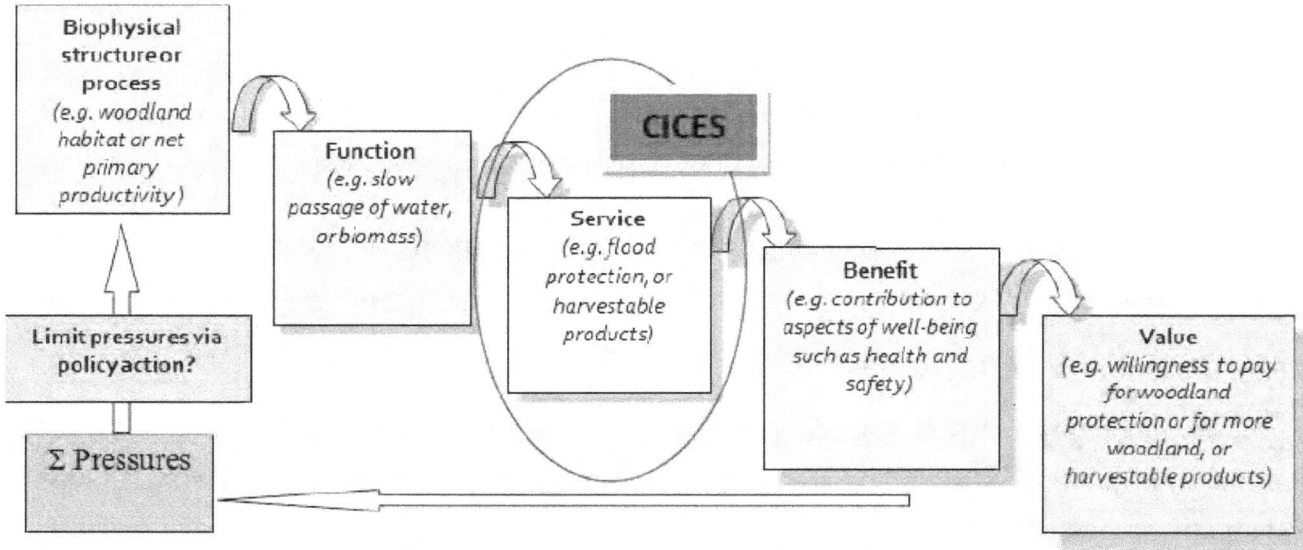

Haines-Young and Potschin (2010a, 2010b, 2013) highlight the problems of using a "flat" classification structure which essentially provides a one-dimensional list of categories, even though they may be grouped in broad types such as regulating, provisioning, cultural, etc. This leads to inflexibility, since the list needs to be updated every time a new service is identified. More importantly, this leads to an unbalanced structure, as the scope of different categories varies. For example, food production and ornamental resources would have the same status under "provisioning," even though the former is of more widespread significance. To avoid these problems, it was proposed that CICES use generic categories and link them in a nested hierarchy to allow for different "scales of concern or thematic content." This hierarchy would also allow for summaries of output at different levels as needed. Thus, the structure of

CICES (Table 2-3) uses generic terminology, which can identify groupings that can progressively be refined according to the interests of the user. Eight categories are proposed, three for provisioning services, three for regulation and maintenance services, and two for cultural services.

Table 2-3. Hierarchical Structure Proposed for CICES

Section	Division	Group
Provisioning	Nutrition	Biomass
		Water
	Materials	Biomass, Fibre
		Water
	Energy	Biomass-based energy sources
		Mechanical energy
Regulation and Maintenance	Mediation of waste, toxics and other nuisances	Mediation by biota
		Mediation by ecosystems
	Mediation of flows	Mass flows
		Liquid flows
		Gaseous/air flows
	Maintenance of physical, chemical, biological conditions	Lifecycle maintenance, habitat and gene pool protection
		Pest and disease control
		Soil formation and composition
		Water conditions
		Atmospheric composition and climate regulation
Cultural	Physical and intellectual interactions with ecosystems and land-/seascapes (environmental settings)	Physical and experiential interactions
		Intellectual and representational interactions
	Spiritual, symbolic, and other interactions with ecosystems and land-/seascapes (environmental settings)	Spiritual and/or emblematic
		Other cultural outputs

Source: Haines-Young and Potschin, 2013. Note that spreadsheets available on the CICES website provide more detailed categories (Class and Class Type), but we do not present them here in the interest of space.

To test whether the data on ecosystem services can be linked to information on economic performance and to ensure that "concrete outcomes" (as described in EPA, 2009) are defined, cross tabulations of CICES groups were done with three international standards for products and activities:

- International Standard Industrial Classification of All Economic Activities (ISIC V4);

- Central Products Classification (CPC); and

- Classification of Individual Consumption by Purpose (COICOP)

Cross-tabulation was also useful in identifying "final outputs" of ecosystems which potentially helps overcome the problem of double counting.[30]

2.8.3 Limitations

Although CICES adapts and expands the MA approach to provide a more systematic and detailed classification system, which includes more attention to the differentiation between intermediate and final ecosystem services, its reliance on the basic MA structure—provisioning, regulating, and cultural service categories—may been seen as a limitation. In particular, this classification approach does not create categories that fully distinguish between (1) what is provided by natural systems, (2) how these natural systems and outputs are used by humans and (3) what is produced by human systems.

2.9   Staub et al. (2011): Indicators for Ecosystem Goods and Services

2.9.1 Objective

Staub et al. (2011), for the Federal Office for the Environment in Switzerland, aim to provide a consolidated inventory of final ecosystem goods and services with concrete proposals for operationalization using indicators.[31] They also develop a methodology for validation and for creating the indicators. Staub et al. consult scientists in specialty research areas as part of the validation process, as well as people in the tourism and nature protection sectors, to verify usability of the inventory, in an attempt to establish relevant baseline data. They develop an inventory of 23 ecosystem services relevant to Switzerland, along with proposals for individual indicators.

2.9.2 Definition and Discussion

In this study, ecosystem services "concentrates on those aspects of ecosystems that have a recognizable connection to (human) welfare, that is, are used or valued in some form or other by the human population" (p. 3). This approach follows Boyd and Banzhaf (2007) in considering only those goods and services that are directly enjoyed, consumed or used by humans as Final Ecosystem Goods and Services (FEGS). It identifies four types of ecosystem goods and services:

---

[30] The ability to link to other ecosystem service classifications such as the MA is also demonstrated in this CICES document.

[31] This report builds on previous efforts to provide "Welfare-significant environmental indicators" (Ott and Staub, 2009)—a new approach for measuring ecosystem services in physical units.

- **Directly usable final ecosystem goods and services:** Used directly by the human population (e.g., recreational or protective services, foodstuffs and feed production, timber yield, contribution to renewable energy);

- **Input factors for market goods:** Not directly consumed (e.g., pollination as an agricultural input);

- **Natural/healthy living environment:** Qualities of health-related environmental media are summarized (e.g., air quality); and

- **Intermediate ecosystem goods and services:** Offer humans no direct benefit. Intermediate ecosystem goods and services are not normally covered here in order to avoid double counting.[32]

The system for identifying these four types is depicted in Figure 2-4. For every FEGS, the benefit it generates (e.g., recreation, prevention, etc.) for the population is also identified. The benefits are assigned to the categories Health, Security, Natural diversity and Production factors to enable establishing links to the product groups used by the Federal Office for the Environment (FOEN). As shown in Figure 2-5, these FEGS can be integrated with MA (2005) and Haines-Young and Potschin (2010a, 2010b, 2013).

To operationalize these FEGS, indicators were developed using the following steps:

- Find the components of nature that generate the goods or services. These are measurable since they are elements of nature (e.g., recreational space for recreational service, protective forests for protection from avalanches, etc.).

- Measure use (demand side) or the supply side of the service. It is important to note that a supply that is not used does not produce any economic benefits.

- Check that there is a connection to welfare.

---

[32] The only exception in the present inventory is CO_2 storage as an input to climate stability. The reasoning here is that the resulting final ecosystem service only emerges after a considerable time delay.

- Specify indicators that can be interpretable without ambiguity. Thus they are selected according to the principle "more is better" (i.e., a higher value of the indicator signifies an increasing amount of the good/service and consequently a higher level of welfare).[33]

- Take into account possibilities for spatial differentiation.

- Determine whether the indicator provides a flow value (benefit contribution per year) rather than a stock value (potential for goods and services).

- Check availability of data

Staub et al. (2011) note the possibility of aggregating to an index relating to the quality of a location or an index relating to health, or even an overall index such as an Ecosystem Services Index (ESI) is possible.

Figure 2-4. System for Dividing the FEGS into the Four Types of Goods and Services (Source: Staub et al., 2011)

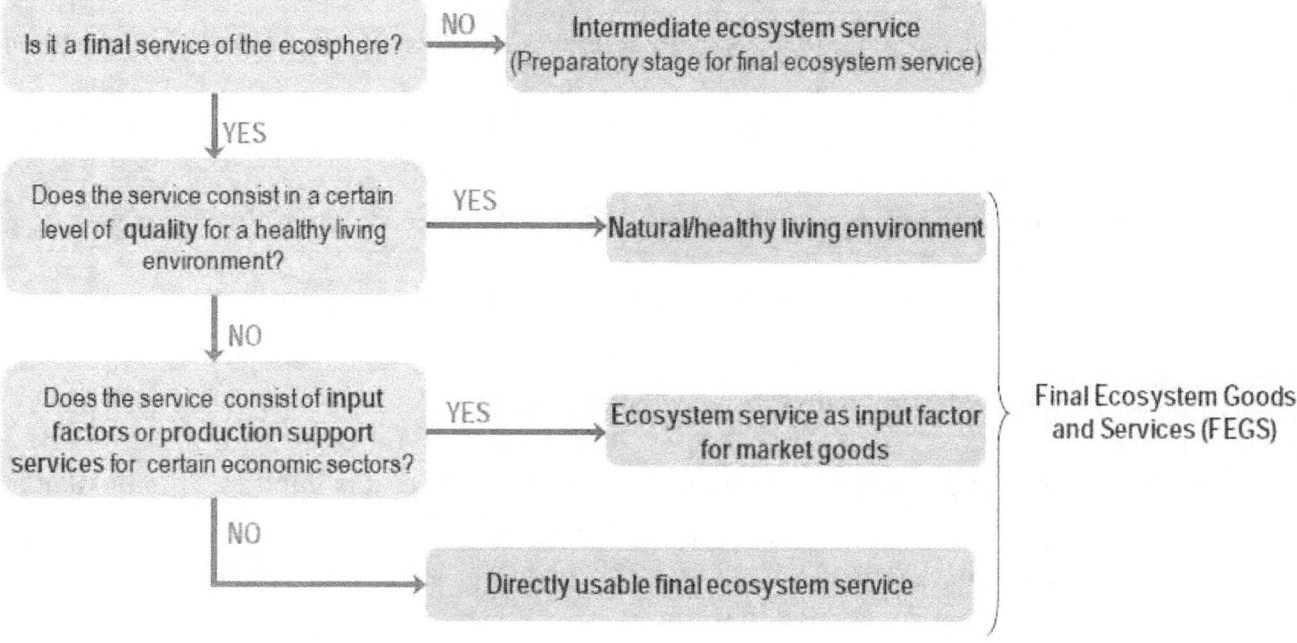

---

[33] It is important to note that this principle only relates to the individual ecosystem service. An overall view is necessary to take into account the change in the overall ecosystem services (e.g., improved access to a recreational area may lead to a decrease in other ecosystem services as a result of an increase in the number of visitors).

Figure 2-5. Integration of the Inventory into the MA and CICES Classifications (Source: Staub et al., 2011)

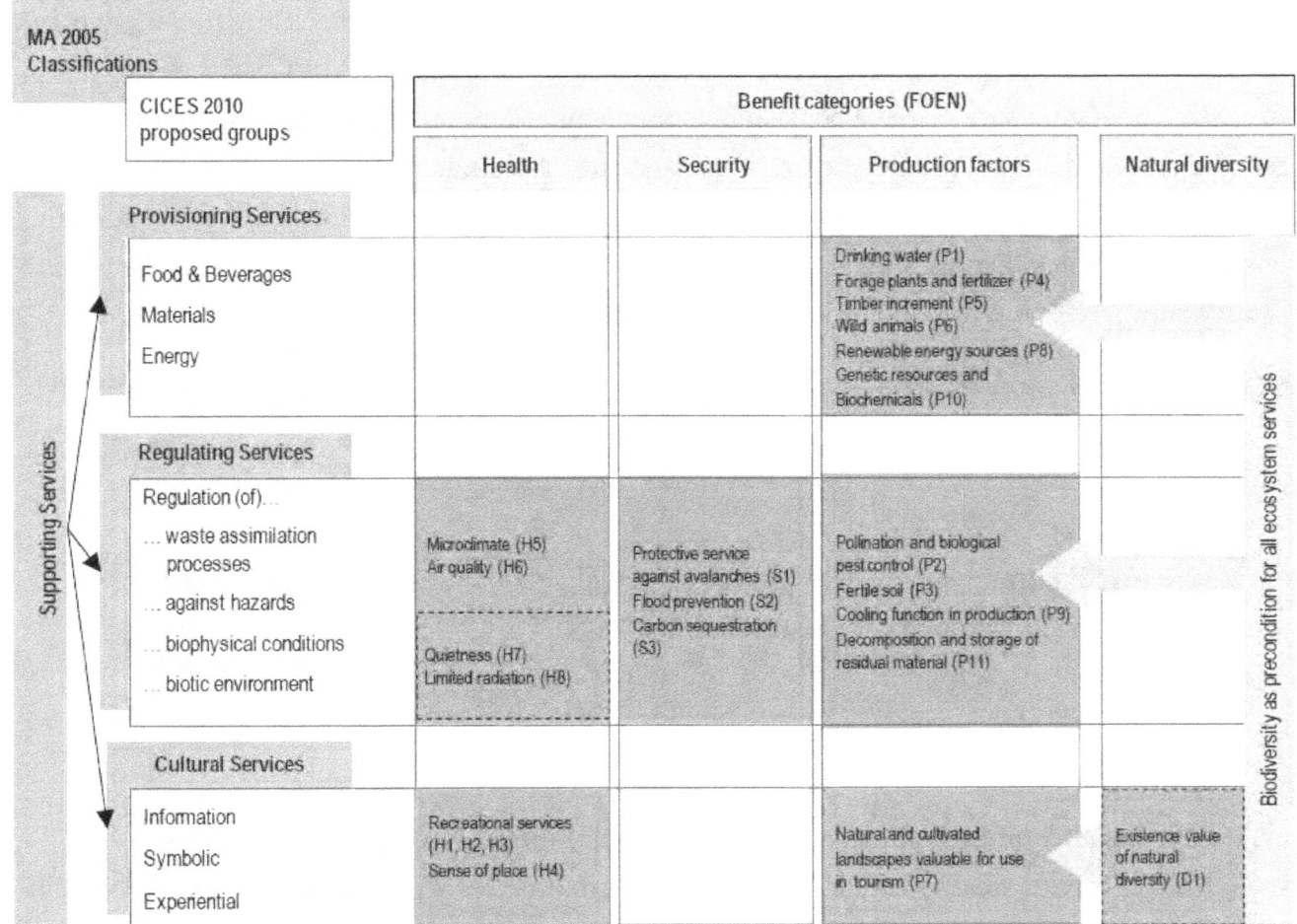

2.9.3 Limitations

Like CICES, the framework proposed by Staub et al. (2011) is fundamentally based on the MA framework which, as described above, has limitations for fully classifying ecosystem services. Also, some of the ecosystem service indicators proposed in the report were specifically based on data available in Switzerland and may not be as relevant or available in other countries.

2.10 Landers and Nahlik (2013): Final Ecosystem Goods and Services Classification System (FEGS-CS)

2.10.1 Objective

This report introduces and describes a detailed classification system for ecosystem services, which focuses specifically on the concept of final ecosystem goods and services

(FEGS).[34] Its definition of FEGS as "the components of nature, directly enjoyed, consumed or used to yield human well-being" is based on the final ecosystem service and ecological endpoint concepts as described by Boyd and Banzhaf (2007) and Boyd (2007). According to the Landers and Nahlik report, the purpose of developing this classification system—FEGS-CS—is to "organize ecosystem services in a consistent and meaningful manner" (p. 15), and to provide "a resource and tool for practitioners to use in consistently defining, identifying, quantifying, and valuing FEGS" (p. 7). It provides an organizing framework that can help to identify the specific ecosystem attributes that are valued by different beneficiary groups, which in turn can be used to identify appropriate metrics and indicators for FEGS.

2.10.2 Definition and Discussion

The FEGS-CS is organized around two main independent classification/categorization components:

1. An "Environmental Class" component, which addresses the question, "Which ecosystems produce ecosystem services?"

2. A "Beneficiary Category" component, which addresses the question, "Who is the beneficiary and what are the FEGS?"

Through the process of answering these two questions, FEGS can be identified or "hypothesized" at the intersection of (or through a combination of) these two main components.

The organizational structure of FEGS-CS is further decomposed by dividing (1) the environmental classes into "environmental subclasses," and (2) the beneficiary categories into "beneficiary subcategories." FEGS-CS also provides a numeric coding system for uniquely identifying individual classes, subclasses, categories, and subcategories, and for defining the hierarchical relationship between the main groupings and their subcomponents.

The environmental class/subclass structure, which is shown in Table 2-4 (along with their numeric codes), is based primarily on the Anderson Land Use and Land Cover Classification system (Anderson et al., 1976).

---

[34] For additional papers and reports related to the development of the FEGS-CS, see Nahlik et al. (2012a), Nahlik et al. (2012b), Ringold et al. (2009), Ringold et al. (2011), Johnston and Russell (2011), and Ringold et al. (2013).

Table 2-4.   FEGS-CS Environmental Classification and Coding

1.	AQUATIC
	11. Rivers and Streams
	12. Wetlands
	13. Lakes and Ponds
	14. Estuaries and Near Coastal and Marine
	15. Open Oceans and Seas
	16. Groundwater
2.	TERRESTRIAL
	21. Forests
	22. Agroecosystems
	23. Created Greenspace
	24. Grasslands
	25. Scrubland / Shrubland
	26. Barren / Rock and Sand
	27. Tundra
	28. Ice and Snow
3.	ATMOSPHERIC
	31. Atmosphere

Source: Landers and Nahlik (2013)

The beneficiary categories and their codes are shown in Table 2-5. FEGS-CS also identifies between one and eight subcategories (along with 2-digit codes) under each beneficiary category (for a total of 38 subcategories). Beneficiaries are defined as "the interests of an individual (i.e., person, organization, household, or firm) that drive active or passive consumption and/or appreciation of ecosystem services resulting in an impact (positive or negative) on their welfare" (p. 13). Because individuals often have multiple interests of this type, each person, household, organization, or firm can be represented in more than one beneficiary category or subcategory. However, the authors emphasize that the categories and subcategories were designed to avoid any duplication of these interests across categories or subcategories, and the report provides written descriptions of each subcategory, to specifically define each one.

The report describes the defined FEGS-CS classification system as an initial structure which "may change as [the authors] further develop and use FEGS-CS." In particular, additional beneficiary groups may be added as the system is used and tested.

Table 2-5.   FEGS-CS Beneficiary Categorization and Coding

XX.XX Beneficiary Categories				
00.01 Agricultural	00.02 Commercial / Industrial	00.03 Government, Municipal, and Residential	00.04 Commercial / Military Transportation	00.05 Subsistence
XX.XXXX Beneficiary Subcategories				
00.0101 Irrigators	00.0201 Food Extractors	00.0301 Municipal Drinking Water Plant Operators	00.0401 Transporters of Goods	00.0501 Water Subsisters
00.0102 CAFO Operators	00.0202 Timber, Fiber, and Ornamental Extractors	00.0302 Waste Water Treatment Plant Operators	00.0402 Transporters of People	00.0502 Food Subsisters
00.0103 Livestock Grazers	00.0203 Industrial Processors	00.0303 Residential Property Owners		00.0503 Timber, Fiber, and Fur / Hide Subsisters
00.0104 Agricultural Processors	00.0204 Industrial Dischargers	00.0304 Military / Coast Guard		00.0504 Building Material Subsisters
00.0105 Aquaculturists	00.0205 Electric and other Energy Generators			
00.0106 Farmers	00.0206 Resource-Dependent Businesses			
00.0107 Foresters	00.0207 Pharmaceutical and Food Supplement Suppliers			
	00.0208 Fur / Hide Trappers and Hunters			
XX.XX Beneficiary Categories				
00.06 Recreational	00.07 Inspirational	00.08 Learning	00.09 Non-Use	00.10 Humanity
XX.XXXX Beneficiary Subcategories				
00.0601 Experiencers and Viewers	00.0701 Spiritual and Ceremonial Participants and Participants of Celebration	00.0801 Educators and Students	00.0901 People Who Care (Existence)	00.1001 All Humans
00.0602 Food Pickers and Gatherers	00.0702 Artists	00.0802 Researchers	00.0902 People Who Care (Option / Bequest)	
00.0603 Hunters				
00.0604 Anglers				
00.0605 Waders, Swimmers, and Divers				
00.0606 Boaters				

Source: Landers and Nahlik (2013)

By combining the two main classification dimensions—environmental classes/subclasses and beneficiary categories/subcategories—the report lays out a detailed set of "FEGS Matrices." These matrices are defined as "a collection of 15 tables that represents the FEGS-CS, in which, for a specific Environmental Subclass, beneficiaries and sets of FEGS are identified and described" (p. 40). Specific FEGS are defined at the intersection of these environmental and beneficiary groupings. Through this combination process, the FEGS-CS report specifically identifies 338 unique FEGS. It notes that this list does not represent an exhaustive catalog of FEGS, rather it provides an initial illustration of the FEGS identification process. In addition, to organize all of these unique FEGS combinations, the report identifies 21 categories of FEGS (Table 2-6).

Table 2-6.   21 FEGS-CS Categories for Organizing FEGS

1	Water
2	Flora
3	Presence of the environment
4	Fauna
5	Fiber
6	Natural materials
7	Open space
8	Viewscapes
9	Sounds and scents
10	Fish
11	Soil
12	Pollinators
13	Depredators and (pest) predators
14	Timber
15	Fungi
16	Substrate
17	Land
18	Air
19	Weather
20	Wind
21	Atmospheric phenomena

Source: Landers and Nahlik (2013)

2.10.3 Limitations

The FEGS-CS directly addresses many of the limitations of the previous approaches by creating a system based on final ecosystem services that differentiates between *what* is provided by natural systems and *how they are used* by humans. However, some of the terminology used in FEGS-CS may present limitations. For example, most of the FEGS-CS categories shown in Table 2-6 can be thought of as stock measures, whereas ecosystem services are inherently more of a flow concept. FEGS-CS does not distinguish between stock and flow concepts. Also the ability to categorize individual people or organizations into multiple beneficiary categories in the FEGS-CS may lead to double counting services if not properly interpreted. For example, the total FEGS received from wild fish stocks by a person who is a recreational angler may include both the specific services received through his/her angling activities as well as from other FEGS (e.g., non-use values). Care must be taken not to double count the other FEGS within the recreational angler beneficiary category.

2.11   Summary

As shown by this review, the existing literature provides a range of definitions and classification approaches for ecosystem services. Table 2-7 summarizes the different objectives and definitions of some of these approaches along with some of their advantages and shortcomings.

In addition to the many differences highlighted above, these classification approaches differ in their treatment or interpretation of the following key concepts:

- natural resource assets and components of nature;

- natural processes/functions;[35]

- service; and

- benefits.

---

[35] Haines-Young and Potschin (2010) distinguish between processes and functions, but most other studies use these terms interchangeably.

Table 2-7. Summary of Ecosystem Services Classification Approaches

Study	Objective	Definition	Advantages	Critiques/Limitations
Daily et al. (1997)	Identify and describe the main connections between ecosystems and human well-being	Conditions and processes through which natural ecosystems help sustain and fulfill human life	Provides an initial list of key examples of how ecosystems help to sustain and fulfill human life	Not a formal classification system. Does not distinguish between ecosystem processes/functions, benefits, and services.
de Groot et al. (2002)	Support comparative ecological economic analyses	Ecosystem functions are 'the capacity of natural processes and components to provide goods and services that satisfy human needs, directly or indirectly"	Clearly delineates 23 categories of ecosystem functions that, indirectly or directly, provide ecosystem services	Risk of double counting by "mixing means with ends" (Wallace, 2007). The term ecosystem functions is unnecessary and confusing (Wallace, 2007; Haines-Young and Potschin, 2010b).
MA (2005)	Provide a link between human welfare and services provided by ecosystems	Ecosystem services are the benefits people obtain from ecosystems	Relatively simple to grasp and apply, includes a wide range of services	Overlap of categories leads to double counting (Fu et al., 2011), "mixing means with ends" (Wallace, 2007), problems of measurement of processes (Boyd and Banzhaf, 2007; Wallace, 2007)
Boyd and Banzhaf (2007)	Develop an accounting system for ecosystem services	Final ecosystem services are components of nature, directly enjoyed, consumed, or used to yield human well-being	Some of the services as defined by them (since they would be stock of fish, etc.) would be much easier to measure than some processes[a]	Functions/processes are also services since they connect ecosystems with welfare. Ecosystem services do not have to be used directly (Fisher and Turner, 2008; Costanza, 2008a) Ecological components do not lead directly to welfare (Fisher and Turner, 2008)[b] Provides conceptual framework and illustration rather than a detailed classification system
Wallace (2007)	Develop a framework for managing landscapes and ecological processes to deliver ecosystem services	Same as MA (2005) except services are defined in terms of structure and components of ecosystems rather than processes	Points out importance of delineating ends and means (Fisher and Turner, 2008)	Does not distinguish between ecosystem services and benefits, and allows for ecosystem services produced with human inputs (Fisher and Turner, 2008)

(continued)

41

Table 2-7.   Summary of Ecosystem Services Classification Approaches (continued)

Study	Objective	Definition	Advantages	Critiques/Limitations
Fisher and Turner (2008)	Answers questions such as: how do ecosystem services deliver human welfare?; where are the benefits realized?; and how do their values change across the landscape in regard to different future scenarios	Ecosystem services are the aspects of ecosystems that are used (either actively or passively) to produce human well-being	Tries to link ecosystems to human welfare while delineating a set of goods—benefits that can be valued	Provides a few examples but not a comprehensive alternative structure for classifying ecosystem services.
Haines-Young and Potschin (2010b)	Provide a framework (CICES) for classifying ecosystem services that is consistent with other international classification systems and allows for linkages with product and activity classifications that form the basis of economic accounting	Contributions that ecosystems make to human well-being, and arise from the interaction of biotic and abiotic processes	Hierarchical structure; clear definition of "pathway" from ecological components to human welfare; ability to link to products and activities used in national income accounting; ability to link to other ecosystem classification systems	Because it uses the MA framework as its basic structure, many of the same potential double counting limitations apply to this approach as well.
Staub et al. (2011)	Develop an inventory of final ecosystem goods and services (relevant to Switzerland) and operationalize them (i.e., provide indicators)	Those aspects of ecosystems that have a recognizable connection to (human) welfare, that is, are used or valued in some form or other by the human population	Provides measurable indicators, easy to link to other classification systems	Many of the limitations of the MA as a classification framework apply to this approach as well. May be difficult to obtain measures in other countries where data are not as readily available.
Landers and Nahlik (2013)	Provide a framework (FEGS-CS) that practitioners can use to define FEGS in a consistent way, to identify relevant ecological metrics, and to move ecosystem services analysis toward quantification and valuation	Based on Boyd and Banzhaf (2007), FEGS are defined as "the components of nature, directly enjoyed, consumed or used to yield human well-being"	Provides detailed hierarchical classification/coding systems for (1) environmental classes, and (2) beneficiary categories. When combined, the two systems can be used to identify unique FEGS	Requires the use of beneficiary categories that do not define mutually exclusive groups of individuals or organizations, but rather mutually exclusive groups of "interests" related to the components of nature

[a] Note that it may be a challenge to measure some services, like "natural surroundings."
[b] Note that Boyd and Banzhaf (2007) indicate that stocks of ecological components may be fairly good proxies of the flow of services from it (if flows are proportional to stocks).

The different treatment of these concepts by different authors hinges on differing opinions as to where ecosystem services are defined to occur along the continuum between ecosystems and human welfare. Some authors view components of nature as the ecosystem service (when consumed directly to yield human well-being) and consider services to be distinct from benefits (e.g., Boyd and Banzhaf, 2007). Others consider the benefits people derive from ecosystems to be ecosystem services (MA, 2005). A related area of disagreement is whether outputs such as recreational angling, which involve human inputs, should be considered to be ecosystem services. Because it is not purely nature's contribution, one viewpoint (e.g., Boyd and Banzhaf, 2007; Fisher and Turner, 2008; Landers and Nahlik, 2013) is that they should be treated as a benefit to which there are important ecological inputs.

There are also disagreements about the treatment of certain natural processes. Some consider natural processes (such as pollination) to be ecosystem services (e.g., Daily, 1997; MA, 2005; Fisher and Turner, 2008), while others consider it inappropriate to include processes as a service. The reasons for not including processes as services include both conceptual problems (people do not care about the actual service, they care about the outcome of the service) and practical implementation problems (e.g., difficulties in measuring processes, lack of sufficient information, difficulty in comparing and trading off other services).

Haines-Young and Potschin (2010a, 2010b, 2013) also provide a very helpful discussion of the distinction between these concepts. However, they also emphasize that whatever terminology is used, a mix of structures, processes, and functions generates the services that ultimately provide benefits to people. Some accounting and assessment studies have adapted some of the classification systems described above to serve their specific objectives. For example, de Groot et al. (2010) adapted MA (2005) to support an assessment of the economic costs of biodiversity loss and ecosystem degradation (*The Economics of Ecosystems and Biodiversity* commonly referred to as TEEB). MA (2005) has also been discussed in private sector guidelines that describe methods to help managers proactively develop strategies to manage business risks and opportunities arising from their company's dependence and impact on ecosystems (e.g., Hanson et al., 2012).

On the other hand, frameworks to assess natural capital (e.g., Porritt, 2007) and environmental accounting systems (e.g., System of integrated Environmental and Economic Accounts [SEEA])[36], have not utilized any existing classification system of ecosystem services to

---

[36]   SEEA (described in more detail in Section 3.2.2) has adopted existing UN systems for classifying economic goods and services but have not extensively applied existing ecosystem service classification systems. In

date. In fact, some studies found that consistent classification systems would be useful in supporting their assessments (e.g., studies that focus on and assess the sustainable use of natural capital such as Maxwell et al., 2014). However, to the best of our knowledge, such studies have not developed any new classification system for ecosystem services and are thus not a part of the literature review in this section.

2.12 Key Lessons Learned

As described in the introduction, the literature on ecosystem services has evolved toward a definition that satisfies four criteria. These include defining services in a way that is consistent and meaningful across different type of services; that is measurable and operational; that is mutually exclusive (to avoid double counting); that matches the context and objective of the study. The major area of disagreement centers on defining the point at which ecosystems deliver services to human beings. How the point of hand-off is defined has crucial implications for developing a classification that meets the four criteria described above.

Moving forward, the review of the literature suggests that certain key features need to be considered in defining and classifying ecosystem services. A brief outline of the key features is provided here.

- **Outlining the objective and context:** As described in the introduction, the analytical objective provides the theoretical underpinning for the definition and classification of ecosystems. Thus, identifying and defining the objective are the first crucial steps in classifying ecosystem services.

- **Defining ecosystem services:** The following discussions are mostly related to what the point of delivery of services should be. How the point of delivery is defined has implications for implementation and/or measurement of services.

  - Define the role of human inputs in producing ecosystem services and benefits: Some studies emphasize that benefits from ecosystem services are realized only when human beings combine natural elements and/or processes with human capital in a household or market production framework. Other studies consider only those components that are purely ecological to be true ecosystem services.

---

addition, although SEEA has developed tables that show physical uses of natural resources by industrial sectors, they have not developed a comprehensive classification system for ecosystem services.

Once human components are included in the picture, the outcomes should be characterized as benefits rather than as services.

- Distinguish between ecosystems components/elements of ecosystems, functions/processes, and benefits: Deciding which of the above should be considered a service, or inputs/precursors to services, is crucial not only in achieving consistency across different services, but also in implementing the definition.

- Distinguish between final and intermediate services: Even though some services may not be directly usable by human beings, but rather "support" or enhance other services, they may provide a service to human beings. The main reason provided for excluding these services from the definition of services is to avoid double counting. It appears, however, that this reason is more of an issue in measuring services than in defining them. It is important to distinguish between intermediate and final services so that services are mutually exclusive; however, there does not appear to be any *a priori* reason for excluding intermediate services from the definition of services. It may be challenging to make this distinction, especially because some services may be intermediate in one context and final in another. For example, clean instream water provides a final service for recreational swimmers but an intermediate service for commercial fishers who rely on clean water to provide healthy fish stocks. Some of the recent literature suggests that identifying existing classes of market products and services that use ecological inputs may be helpful in defining what "final outcomes" of ecosystem services are.

- Identify beneficiaries: Services are only valuable if human beings perceive them to be valuable and/or use them. Also, different groups can derive different services from the same ecological resources. Thus, identifying the different ways in which humans use and benefit from the resource is key to defining something as an ecosystem service. Classification systems for economic goods and services also consider similarities in human uses when grouping products; however, one important difference is that market products are often designed to meet specific human needs or uses, whereas ecosystem services are not.

- **Measuring services:** An interesting issue raised in some studies is whether we need to measure a service, or whether it would be sufficient to measure the value of a service.

The answer depends on the relevant context/objective. Measuring values may be appropriate for some natural resource management decisions or cost-benefit analysis, but not for other contexts. Other issues to keep in mind are as follows:

- Consider whether a flow or a stock is appropriate: Measures of both stocks and flows may be relevant for certain types of services because they may be indicative of the current and the future potential for ecosystems to provide services. For other services, flows may not be appropriate measures.

- Use proxies or indicators: Perfect measures of ecosystem services are rarely readily available. However, proxies or indicators may be available, and these may be sufficient – for example, in accounting frameworks, as measures of the condition and potential of ecosystems to provide services.

- Consider spatial and temporal aspects: Most of the literature agrees that ecosystem services and the values associated with them vary substantially both spatially and temporally. Not only are these tied to the culture of the local population, but they are also very closely tied to the surrounding land use and land cover. They also vary temporally because of both human influence and natural evolution.

- Identify datasets: Although this step has not been emphasized in most of the literature, this is a key step to operationalizing definitions (Staub et al., 2011).

- **Classifying ecosystem services:** The more recent literature suggests grouping according to benefit categories that are also consistent with product categories. This helps provide a framework that is both useful for accounting and for cost-benefit analysis (by helping identify what "final outcomes" are). Using a flexible nested hierarchy also has certain merits such as ease of adding services and summarizing and/or analyzing data at different levels.

SECTION 3
REVIEW OF ECONOMIC CLASSIFICATION AND ACCOUNTING SYSTEMS

3.1 Introduction

To develop a conceptual framework and classification system for ecosystem services (NESCS), we draw on the concepts and methods used for classifying economic goods and services. Although there are differences between traditional income and product accounting and ecosystem service accounting, many of the underlying principles for economic classification can be applied in developing NESCS. In Section 3.2, we review the main systems used to organize, compile, and report economic accounts. In Section 3.3, we describe the main classification systems used to support these accounts. We focus on the elements and features of these systems that are most relevant for developing NESCS. In Section 3.4, we discuss the implications for the design of NESCS.

3.2 What Are the Main Systems of Economic Accounts?

Not surprisingly, most systems of national accounts focus on economic activity within the market sector of the economy. Therefore, we begin by reviewing the main systems used for market-sector accounting; however, we also discuss how these systems are being adapted and expanded to include non-market sectors and activities.

3.2.1 National Market-Sector Accounts

The **National Economic Accounts** (NEA) are the main system of accounts used in the United States to measure national market-based economic activity in the country. The Bureau of Economic Analysis (BEA), an agency under the U.S. Department of Commerce, has primary responsibility for providing these accounts. The NEA are designed to answer two fundamental questions: (a) what is the output of the economy (size, composition, and use), and (b) by what economic process is this output produced and distributed? The two main components of the NEA used to address these questions are:

- National Income and Product Accounts (NIPAs); and

- Input-Output (I-O) Accounts.

In addition, the NEA includes capital finance ("flow-of-funds") accounts that track monetary and credit transactions in the economy.

Through the United Nations (UN), the international community (including the United States) has also developed the System of National Accounts (SNA), which provides an internationally accepted set of guidelines for compiling national accounts.[37] As feasible, to maintain consistency between systems, the BEA incorporates the SNA guidelines into the development of the NEA; however, some differences persist (Mead et al., 2004).

3.2.1.1 The National Income and Product Accounts (NIPAs)

The NIPAs consist of seven main accounts designed to provide a consistent and comprehensive picture of production, distribution, consumption, investment, and savings in the economy. Of these seven accounts, the one that receives the most attention and interest is the domestic income and product account, which includes measures of **gross domestic product** (GDP) (BEA, 2009).

GDP for the U.S. economy is defined as a measure of "the value of final goods and services produced in the United States in a given period of time" (BEA, 2007).[38] Several points are worth noting about this measure. First, it primarily includes **market** goods and services involving economic transactions.[39] Second, it is valued using market prices for the goods and services. Third, it measures the value of **final** goods and services (i.e., those intended for end users and not as inputs into additional stages of production). The value of intermediate products, which are used as inputs for other products, is assumed to be included within the price of these final products. Fourth, it is a **flow** measure, capturing economic activity over a period of time, rather than a stock measure (e.g., capital equipment or inventory levels) at a point in time.

3.2.1.2 I-O Accounts

The I-O accounts serve as both the data source and the framework for preparing national income accounts. Unlike the NIPAs, they trace the flow of goods and services between industries in the production process (i.e., they include *intermediate* goods and services), and they represent the *value added* by each industry. They show the flow of goods and services from each industry to other industries and to final users in the economy. The benchmark accounts,[40] which are produced every 5 years, include information on more than 425 industries. The annual I-O accounts include information on 65 industries.

---

[37] The most recent guidelines are laid out in the "SNA 2008" report (Commission of the European Communities, International Monetary Fund, Organization for Economic Cooperation and Development, United Nations and World Bank. 2009. *System of National Accounts, 2008* (United Nations publication, Sales No. E.08.XVII.29).

[38] In contrast, **gross national product** (GNP) measures the output produced by a country's residents, irrespective of where it is produced.

[39] Exceptions include, for example, the services provided by owner-occupied housing.

[40] These accounts determine the structure and level of GDP for comprehensive revisions of the national income and product accounts.

3.2.2 Non-market Accounts

In many instances, activity occurring outside the traditional market sector can have important implications for economic productivity and well-being. Below we discuss how both macro- and micro-level accounting systems can and are being adapted to address non-market elements, in particular the contribution of natural and environmental systems.

3.2.2.1 National Environmental and Economic Accounts

Although the NEA and GDP (in particular) measures provide useful indicators of the state of the economy, they are often criticized for being incomplete. The primary reason for this criticism is the exclusion of non-market features in or affecting the economy, such as unpaid work, leisure activities, investment in human capital, household health production, and the environment (Nordhaus, 2000). As a result, many experts have argued in favor of a system of "satellite accounts." These accounts are not intended to replace the NIPAs but rather to serve as a complementary system that provides a more comprehensive picture of the economy and human well-being (see, for example, NRC, 2005).

Satellite accounts for environmental and natural resources are among the most commonly discussed and investigated types of non-market accounts. As for the market-based SNAs, the international community has developed a set of internationally accepted guidelines for creating these accounts. In 2003, the UN issued a handbook for what is called the System of integrated Environmental and Economic Accounts (SEEA). This document proposes the following four main categories of satellite accounts:

1. Physical flow accounts, including the flow of natural materials (e.g., wood), resources, and energy as they relate to goods and services produced in the economy, and hybrid accounts that combine the physical flows with related economic input data for production activities;

2. Economic accounts and environmental transactions, including expenditures made by businesses, governments, and households to protect the environment;

3. Environmental asset accounts, which measure stocks in physical and monetary terms, for example, timber stock accounts showing opening and closing timber balances and the related changes over the course of an accounting period; and

4. SNA extension accounts to incorporate natural resource depletion, degradation, and defensive expenditures.

The SEEA-2003 framework is currently being applied in three main areas: water, energy, and land and ecosystems. All three areas require accounting for ecosystem services, in particular the land and ecosystems category, which is currently under development and being coordinated by the World Bank through the Global Partnership on Wealth Accounting and the Valuation of Ecosystem Services (WAVES, 2013).

At this stage, the framework for water accounts—SEEA-Water (UN, 2012)—is the best defined of the three areas. This framework includes the following main components:

- flow accounts, including physical water supply and use tables, water pollutant emission accounts, and hybrid physical-economic accounts;

- asset accounts, including water quantity and water quality accounts; and

- valuation of non-market water resources.

The flow tables account for non-market water services by quantifying the physical flows of abstracted water to different sectors of the economy (according to different types of uses). The valuation component is intended to convert those non-market flows into monetary terms; however, this account is considered "experimental" and has not yet been implemented in a country-wide application.

One of the potential applications of the SEEA is to support the development and estimation of **green GDP**. Although there are different interpretations of this concept, the fundamental idea is to address the perceived limitations of traditional GDP measures by (1) deducting annual losses associated with natural resource depletion and environmental pollution (Wu and Wu, 2010), or (2) expanding traditional GDP to separately account for the non-market public good benefits provided by nature each year (Boyd, 2007).

3.2.2.2 Micro-level Environmental Accounts

In addition to adapting national-level accounts, there is increasing interest in and movement toward adapting private sector corporate accounts to incorporate environmental factors (Waage and Kester, 2013). These changes involve at least two types of adjustments:

- accounting for the role and contribution of natural resources and ecosystem services in corporate performance; and

- measuring the impacts of private-sector activities on ecosystem services to the public.

50

Examples of other accounting/assessment work done to address issues of sustainability of natural capital include Hanson et al. (2012) and Porritt (2007). However, in contrast to national accounts, no commonly agreed upon framework like the SEEA has emerged to address these changes. Instead, companies and industry groups have tended to work independently to address their specific needs and requirements. For example, the oil and gas industry has developed a checklist system to identify specific dependencies and impacts on ecosystem services associated with specific industry activities in specific habitats (International Petroleum Industry Environmental Conservation Association [IPIECA], 2011). This checklist system does not yet quantify the dependencies or impacts of ecosystem services, but it is a first step in that direction.

3.3   How Are Classification Systems Used in Economic Accounts?

For several reasons, classification systems play a critical role in the development and reporting of economic accounts. Below we review the main types, features, and roles of existing economic classification systems.

3.3.1   North American Classification Systems

The BEA currently uses two main alternative classification systems in the NEA to categorize and account for the flow of market goods and services in the economy:

- the North American Industry Classification System (NAICS); and

- the North American Product Classification System (NAPCS).

Both systems were developed jointly by Canada, the United States, and Mexico, to allow for high-level comparability of business statistics among the North American countries.

NAICS is the standard used by federal statistical agencies in classifying business establishments for the purpose of collecting, analyzing, and publishing statistical data related to the U.S. business economy. It was adopted in 1997 to replace the Standard Industrial Classification (SIC) system. Due to its focus on the producers of economic goods and services, it can be thought of as a "supply-side" classification system.

NAPCS is the standard currently being developed to classify all of the products being produced by business establishments in the economy. The two main objectives of NAPCS are:

1. to "identify, define, and classify the final outputs (outputs regardless of their designation as intermediate or final demand) produced and transacted (sold

transferred, or placed in inventory) by the reporting units within each industry" (ECPC, 2003, p. 2); and

2. to "develop a demand-based, hierarchical aggregation system, in which products are grouped according to *how they are principally used* [emphasis added] and according to how they are used in relationship to each other in satisfying that principal use" (ECPC, 2003, p. 2).

Although it is fundamentally a "demand-side" system, NAPCS is also being developed so that it can be linked to the NAICS industry structure.[41]

3.3.2   UN Classification Systems

Through the UN, the international community has also developed classification systems for industries and products. For supply-side classification, the UN uses the International Standard Industrial Classification (ISIC) system, which classifies establishments in a way that is very similar to NAICS (UN, 2003). For demand-side classification, they have developed the Central Product Classification (CPC) system. Like NAPCS, CPC provides a product classification system for both goods and services; however, NAPCS was developed in part to address perceived shortcomings of CPC. In particular, CPC was "not based on a single unifying theoretical principle for grouping and aggregating products." (Mohr, 2002, p. 3).

3.3.3   The Role of NAICS and NAPCS in U.S. Economic Accounts

The NAICS system for classifying industries and establishments was developed in part to allow analysts to track the flow of intermediate goods and services between industrial sectors (as part of the I-O accounts) and to distinguish them from final goods and services sold to consumers. The NAICS coding system involves a six-digit hierarchical structure, where the first two digits designate the general sector, and each subsequent digit represents a more detailed subset of the sector or industry. Table 3-1 provides a list of the 20 major (two-digit) NAICS sectors. Note that sectors 11 to 45 primarily involve production of goods whereas the remaining sectors primarily produce services.

---

[41] For example, a flu shot can be provided by a doctor's office, a hospital, or a walk-in clinic. These three units are classified to three different NAICS industries. If data users want information about all flu shots provided, they would need to be able to identify and aggregate the individual products coming out of the three different industries. Thus, in many cases, the need for specific statistical data is better addressed with product data crossing industries rather than with the creation of a new industry. This is particularly true with NAICS, which groups establishments into industries based on their production function. For more details, see http://www.census.gov/eos/www/naics/reference_files_tools/NAICS_Update_Process_Fact_Sheet.pdf.

Table 3-1.   2012 2-Digit NAICS Codes and Sectors

NAICS Code	NAICS Sector
11	Agriculture, Forestry, Fishing and Hunting
21	Mining
22	Utilities
23	Construction
31–33	Manufacturing
42	Wholesale Trade
44–45	Retail Trade
48–49	Transportation and Warehousing
51	Information
52	Finance and Insurance
53	Real Estate Rental and Leasing
54	Professional, Scientific, and Technical Services
55	Management of Companies and Enterprises
56	Administrative and Support and Waste Management and Remediation Services
61	Educational Services
62	Health Care and Social Assistance
71	Arts, Entertainment, and Recreation
72	Accommodation and Food Services
81	Other Services (except Public Administration)
92	Public Administration

Source: U.S. Census Bureau (http://www.census.gov/eos/www/naics/index.html, accessed May 29, 2015)

Focusing on the supply-side of the economy, the NAICS classification is designed to group together entities engaged in similar production activities. BEA typically refers to these production entities as "establishments,"[42] which it defines as "a single physical location where business is conducted, or where services are performed." Conceptually, this approach means grouping establishments with similar production functions. Accordingly, each establishment in the economy is assigned to a single primary NAICS code, which corresponds to its primary production activity (recognizing that some establishments produce multiple products and therefore also have secondary activities).

To provide a demand-side counterpart to the NAICS, the North American Free Trade Agreement (NAFTA) countries have begun to develop the NAPCS. This classification system is intended to categorize commodities rather than establishments, where the categories reflect

---

[42] Examples of an establishment include a factory, mill, store, hotel, movie theater, mine, farm, airline terminal, sales office, warehouse, or central administrative office.

similar product characteristics rather than production processes. In this system, each commodity produced and sold in the economy should be assigned to a single NAPCS code.

According to the Economic Classification Policy Committee of the United States (ECPC, 2003), one of the broad main objectives of the NAPCS is to create a system where "products are grouped according to *how they are principally used* and according to how they are used in relationship to each other in satisfying that principal use" [emphasis added]. This objective entails grouping products that are close substitutes, either as inputs to production processes or as inputs to consumption.[43] Thus NAPCS provides a demand-side perspective that focuses on the similarities in how products are used, either by firms or by households, and NAICS provides a supply-side perspective that focuses on similarities in how products are produced. Table 3-2 shows a few examples of product groups in NAPCS, as reported in a recent provisional classification structure developed by Statistics Canada. Although preliminary, this demand-side list provides a useful contrast to the supply-side NAICS groupings in Table 3-1.

Table 3-3 further highlights the differences between NAICS and NAPCS using two examples. Example 1 shows how NAICS subdivides the electricity generating sector according to differences in production processes, whereas NAPCS only includes one product category for electricity because it's use does not depend on how it is produced. This example shows how a single product category can be produced in more than one industry sector. In contrast, Example 2 shows how, from a production process perspective, NAICS treats "Other Basic Inorganic Chemical Manufacturing" as a single sector, whereas NAPCS subdivides the products from this sector into multiple chemical categories based on differences in their fundamental uses. This example shows how a single sector can produce multiple products.[44]

One longer-term objective of the combined NAICS-NAPCS systems will be to provide a detailed "cross-walk" between individual NAPCS product categories and the possibly multiple NAICS industry categories that produce them (and vice versa).[45]

---

[43] Technically, if two products are "substitutes," then if the price and/or scarcity of one product increases, it will increase the demand of the other (substitute) product. If they are "complements," the opposite will occur.

[44] Both NAICS and NAPCS use hierarchical coding systems, with additional digits representing additional levels of hierarchical subdivisions. When there is only one element in a group it is assigned both a low and high level code. For example, in NAPCS, the codes 145, 14511, and 145111 all refer to "electricity" since it is the only product in the 3-digit category.

[45] A preliminary cross-walk is available at http://www.census.gov/eos/www/napcs/ (accessed May 29, 2015).

Table 3-2. Selected NAPCS Canada 2012 3-Digit Codes and Groups

Code	High-Level Product Groups
111	Live animals
112	Wheat
121	Fish, shellfish and other fishery products
131	Logs, pulpwood and other forestry products
142	Natural gas
161	Potash
172	Meat products
193	Bottled water, carbonated soft drinks, other beverages, and ice
231	Clothing, footwear and accessories
271	Basic chemicals
371	Electronic and electrical parts
412	Medium and heavy trucks, buses and other motor vehicles
511	Transportation of commodities by pipeline
541	Warehousing and storage services
581	Rental and leasing (except rental of real estate)
712	Advertising, public relations, and related services
831	Sport and live performance services
871	Public administration services

Source: Statistics Canada; North American Product Classification System (NAPCS) Canada 2012
http://www.statcan.gc.ca/subjects-sujets/standard-norme/napcs-scpan/2012/index-indexe-eng.htm (accessed
May 29, 2015).

3.3.3.1 Key Features of the NAICS and NAPCS Classification Systems

The NAICS and NAPCS systems contain several features that make them particularly useful for national economic accounting. These features including the following:

1. they provide organizing structures that facilitate a wide range of accounting activities, including the collection, tabulation, presentation, and analysis of economic data;

2. they provide consistent systems that promote uniformity of data and allow for more accurate comparisons of economic activity across sectors and over periods of time;

3. by providing hierarchical classification systems, they help in presenting accounts at different levels of aggregation; and

4. by providing adaptable, nested classification structures, they offer flexibility for adding to and expanding the number of subcategories as needed over time.

These features make them particularly well suited for helping to track the flow of goods and services within the economy, and between sectors through the I-O accounts.

Table 3-3. NAICS-NAPCS Comparison

	Example 1			
NAICS			**NAPCS**	
Code	**Description**	**Code**		**Description**
22111	Electric Power Generation	145	Electricity	
221111	Hydroelectric Power Generation	14511	Electricity	
221112	Fossil Fuel Electric Power Generation	145111	Electricity	
221113	Nuclear Electric Power Generation			
221114	Solar Electric Power Generation			
221115	Wind Electric Power Generation			
221116	Geothermal Electric Power Generation			
221117	Biomass Electric Power Generation			
221118	Other Electric Power Generation			

	Example 2		
NAICS		**NAPCS**	
Code	**Description**	**Code**	**Description**
32518	Other Basic Inorganic Chemical Manufacturing	27112	Other Basic Inorganic Chemicals
325180	Other Basic Inorganic Chemical Manufacturing	271121	Sulfuric acid
		271122	Chlorine
		271123	Sodium hydroxide
		271124	Inorganic potassium and sodium compounds
		271125	Carbon black
		271126	Chemical catalytic preparations
		271127	Nuclear fuel
		271128	Other

3.3.4   Relationships Between NAPCS and NAICS in Economic Accounts: Input-Output Framework

In this section, we describe the accounting structure that links the NAICS and NAPCS systems. The national I-O accounts are comprised of two main tables for tracking the flow of goods and services within the economy.

1. a **make table**, which displays the production of different commodities across industries; and

2. a **use table**, which displays the use of commodities across different intermediate and final users.

Table 3-4 provides a simplified example of a summary level I-O make table for the economy. The NAICS industry categories are separated into separate columns and the NAPCS commodity categories in rows.[46] The purpose of the table is to show how the production (make) of each commodity is divided across the different productive sectors of the economy. It shows that, although the NAICS and NAPCS classifications have similar structures and category names (e.g., both contain educational services as a separate category), there is not a one-to-one relationship between them. Many categories of commodities in NAPCS can be produced by more than one NAICS industry sector (e.g., food can be from agricultural and manufacturing sectors), and many industries can produce more than one commodity (e.g., manufacturing produces clothing and automobiles). Other examples are shown in Table 3-4.

In addition, Table 3-4 requires that the individual NAPCS and NAICS categories represent mutually exclusive commodity and industry categories, respectively. In other words, each commodity must be associated with a single NAPCS code, and each establishment must be associated with a single primary NAICS code. This condition ensures that products are not double-counted in the accounting framework. Each cell in the table represents the total dollar amount of the product (row) that is produced by the industry category (column) during a specific time period. The total value of production in each industry NAICS category (column) can be generated by adding up the values in each NAPCS commodity category (row). Similarly, the total production value for each commodity can be generated by summing across industry categories.

---

[46] Because the NAPCS is still in development, the *actual* I-O make and use tables for the United States reported by the Bureau of Economic Analysis (http://www.bea.gov/industry/iedguide.htm#io) use a modified version of the NAICS categories for both the industry and commodity groupings. However, once the NAPCS is completed, the plan is to use it as the basis for commodity classification (BEA, 2009; http://www.bea.gov/papers/pdf/IOmanual_092906.pdf).

Table 3–4. Example of I-O Make Table Relating NAPCS[a] and NAICS Categories

Addresses the question: which industries (columns) make which commodities (rows)?

	Industries (producing a commodity) →													
Commodities (produced by an industry) ↓	Agriculture, Forestry, Fishing And Hunting	Mining	Utilities	Construction	Manufacturing	Wholesale and Retail Trade	Transportation and Warehousing	Information	Finance and Insurance	...	Educational Services	...	Public Administration	Total Commodity Output
NAICS Industry code	11	21	22	23	31–33	42–45	48–49	51	52	...	61	...	92	
111 Live animals														
112 Wheat														
121 Fish, shellfish and other fishery products														
131 Logs, pulpwood and other forestry products														
142 Natural gas														
161 Potash														
172 Meat products														
193 Bottled water, carbonated soft drinks, other beverages, and ice														
231 Clothing, footwear and accessories														
271 Basic chemicals														
371 Electronic and electrical parts														
412 Medium and heavy trucks, buses and other motor vehicles														
511 Transportation of commodities by pipeline														
541 Warehousing and storage services														
581 Rental and leasing (except rental of real estate)														
712 Advertising, public relations, and related services														
761 Financial services (except insurance)														
782 Water, sewer, and waste management services[b]														
821 Education Services														
831 Sport and live performance services														
871 Public administration services														
....														
Total industry output														

NAPCS commodity code

[a] Only selected NAPCS categories are shown in this table. Source of NAPCS categories: Statistics Canada 2012
[b] Includes environmental remediation services in 78231 (Waste management and remediation services)

58

Table 3-5 provides a simplified example of a summary level I-O use table for the economy. Similar to the make table, the NAPCS commodity categories are separated into rows. However, in this table the columns represent categories of commodity users, rather than commodity producers. The purpose of the table is to show how spending on (use of) each commodity is divided across different sectors of the economy. Despite this different objective, most of the column categories are the same as in the make table. They mainly include industry sectors, but in this table, these sectors are also combined to represent total "intermediate" uses of the commodities. The additional columns, which make up the "final" uses, mainly include a category for purchases by the household sector. As in the make table, there is not a one-to-one correspondence between the commodity and sector categories. Many sectors can use the same commodity category, and many commodity categories can be used by the same sector.

To avoid double counting, Table 3-5 also requires that the individual NAPCS and NAICS categories represent mutually exclusive commodity and industry categories. The total value of goods and services purchased by each NAICS category (and by households) can be generated by adding up across NAPCS commodity category rows. Similarly, the total spending on each commodity can be generated by summing across columns. In this table, each cell represents the total dollar amount of the product category (row) that is purchased by the industry or other user category (column) during a specific time period.

Table 3-5. Example of I-O Use Table Relating NAPCS[a] and NAICS Categories

Addresses the question: which commodities (rows) are used by which sectors/users (columns)?

NAPCS commodity code	Commodities (used by sectors) ↓ / Sectors (using a commodity) →	Agriculture, Forestry, Fishing And Hunting	Mining	Utilities	Construction	Manufacturing	Wholesale and Retail Trade	Transportation and Warehousing	Information	Finance and Insurance	...	Educational Services	...	Public Administration	Total intermediate use	Household personal consumption expenditures	Other	Total final uses	Total commodity use
	NAICS Industry code	11	21	22	23	31–33	42–45	48–49	51	52	...	61	...	92					
111	Live animals																		
112	Wheat																		
121	Fish, shellfish and other fishery products																		
131	Logs, pulpwood and other forestry products																		
142	Natural gas																		
161	Potash																		
172	Meat products																		
193	Bottled water, carbonated soft drinks, other beverages, and ice																		
231	Clothing, footwear and accessories																		
271	Basic chemicals																		
371	Electronic and electrical parts																		
412	Medium and heavy trucks, buses and other motor vehicles																		
511	Transportation of commodities by pipeline																		
541	Warehousing and storage services																		
581	Rental and leasing (except rental of real estate)																		
712	Advertising, public relations, and related services																		
761	Financial services (except insurance)																		
782	Water, sewer, and waste management services[b]																		
821	Education Services																		
831	Sport and live performance services																		
871	Public administration services																		
....																			
	Total sector use																		

[a] Only selected NAPCS categories are shown in this table. Source of NAPCS categories: Statistics Canada 2012

[b] Includes environmental remediation services in 78231 (Waste management and remediation services)

3.4　Implications of the NAICS and NAPCS Systems for Developing NESCS

Many of the key features and principles used to develop the NAICS and NAPCS systems are also relevant for conceptualizing and developing NESCS. Like NAICS and NAPCS, NESCS should provide (1) an organizing structure that facilitates accounting activities, in part by specifying mutually exclusive categories; (2) a consistent system that promotes uniformity of data collection and development; (3) a hierarchical classification structure that allows for accounts at different levels of aggregation; and (4) a flexible structure suitable for adding to and expanding subcategories as needed.

In addition, the important distinction between supply-side and demand-side classification systems in NAICS and NAPCS can also be applied to ecosystem services. Just as economic goods and services can be classified in NAICS according to supply-side characteristics and in NAPCS according to demand-side characteristics, ecosystem services can also be classified according to both supply- and demand-side characteristics. However, because ecosystem services are produced only by natural systems, the supply-side classification for ecosystem services must focus on natural systems rather than on human production systems. On the demand side, ecosystem services must be grouped according to how (and by whom) these natural systems are used and enjoyed by humans.

Despite these connections, there are also important differences between economic classification and ecosystem services classification. First, whereas economic classification systems can use information provided by market transactions to define categories of goods and service, the non-market nature of ecosystem services and the resulting lack transaction information make them more challenging to define and categorize.[47] In most instances, market transactions records (e.g., receipts or invoices) describe the type of product (good or service) being exchanged for money. Developing a classification system for market products (i.e., NAPCS) is therefore less about *defining* what is exchanged than it is about *grouping* similar products. Moreover, the availability of price data makes it feasible in some cases to group market products based on similarities in their prices and their estimated cross-price elasticities (ECPC, 1994). In contrast, to develop a classification system for non-market products, one must first define the implicit "commodities" that are being exchanged. Defining service commodities is inherently challenging because they tend to be less tangible than goods. The absence of transaction data for ecosystem services makes this process even more difficult.

---

[47]　ECPC (2001) includes some initial discussions on how to incorporate non-market goods and services into economic classification; however, they are still largely excluded from NAICS and NAPCS.

Second, the ability in NAICS to classify all economic production establishments into individual and mutually exclusive categories is made possible largely because market systems encourage producers to specialize. This specialization allows NAICS to classify producers according to their *primary* production activity. The same incentives for specialization do not exist for ecosystems; therefore, it is much less meaningful to develop a classification system that categorizes ecosystems according to their primary ecological production processes. As a result, alternative approaches are needed to classify the supply side of ecosystem services.

Third, the role and importance of categorizing human uses is different for ecosystem service classification than for economic classification. One main reason for this difference is that economic products are often designed to meet specific human needs and requirements. For example, automotive manufacturers produce vehicles with different attributes (e.g., size, seating, power, fuel efficiency) to satisfy different uses (e.g., commuting, recreation, passenger transport, transport of goods and equipment). As a result, categorizing these products according to their physical attributes is often equivalent to categorizing them according to their uses. In contrast, ecosystems do not produce outputs that are designed for specific human use. In many cases they can support a multiplicity of human uses with a wide range of benefits and values. For example, migratory birds can provide hunting and bird watching benefits for recreators and pest control benefits for some farmers. Each use reflects a different service with potentially very different values. Moreover, in many cases the individual uses do not rival each other and can all be enjoyed (i.e., there are public good benefits). Therefore, categorizing ecosystem services requires not only an understanding of *what* is provided by nature but also *how* it may be used in different ways.

The availability of price data for economic goods and services also reduces the need to classify them according to how they are used. Although users may have different values for the commodity, at the margin their values should approach the market price which is the same for everyone.[48] Moreover, those who value the commodity less than its price do not purchase or use it. Therefore, price information can substitute to some extent for user information in classifying market goods and services. This type of price information is not available for non-market goods and services. Therefore, when classifying ecosystem services, the absence of a price makes it more important to consider the different ways in which natural systems are used.

---

[48] For example, rental rates for sports utility vehicles are generally the same, regardless of differences in how renters use them (e.g., recreation/tourism, transporting equipment, or transporting passengers). NAICS and NAPCS do not specify separate commodity categories based on how the rental vehicles are used.

There is also a fundamental difference between the main analytical objectives of NAICS/NAPCS and NESCS. NAICS and NAPCS were mainly designed to support the development of national economic accounts, which measure trends in market-based economic activity across, time, sectors, and regions. In contrast, the main objective of NESCS is to support analyses of how policy-induced changes in ecosystems affect human well-being (e.g., cost-benefit analysis).

Despite these differences, there are also important areas of overlap between these two analytical objectives. First, as discussed in the next section, national income accounting and cost-benefit analysis are based on the same underlying conceptual framework for economic analysis. Second, by expanding this framework to include non-market ecosystem services, the development of NESCS may provide a classification structure that is also applicable for constructing green GDP and other national environmental accounts.

SECTION 4
NESCS CONCEPTUAL FRAMEWORK, CLASSIFICATION STRUCTURE, AND CODING SYSTEM

4.1   Introduction

The main goal of NESCS is to help identify the distinct "pathways" through which policy-induced changes in ecosystems ultimately lead to changes in human welfare. To do this, we begin by defining a conceptual framework for linking the ecological systems that produce ecosystem services with the human systems that directly use them (i.e., market production systems and households). Section 4.2 describes the conceptual framework for NESCS, which draws from and adapts concepts underlying the economic accounting and classification systems.

Using this framework we then define the NESCS. This system is designed to provide an exhaustive set of mutually exclusive categories for linking ecosystem outcomes to direct human uses. Again, building on the concepts from economic concepts and principle, we argue that services are defined as flows from the producers/providers to consumers/users. Thus, in order to identify and define FFES, we first need to identify producers (or "supply-side") and consumers (or "demand-side") of the service. As described in Section 3, the two existing classification systems for economic goods and services in the United States (North American Industry Classification System (NAICS) and North American Product Classification System (NAPCS) also distinguish between "supply-side" and "demand-side" systems. To reiterate, the NAICS system is designed to classify the production processes for goods and services based on a supply-side perspective (i.e., who is producing the commodities and how), whereas the NAPCS system focuses on the demand-side perspective to classify the goods and services (i.e., how and by whom are the products being used). NESCS also distinguishes between a supply-side grouping and a demand-side grouping. We thus include two complementary components in the NESCS architecture:

- NESCS-S, which refers to the supply-side classification of ecosystems and ecological end-products; and

- NESCS-D, which refers to the demand-side classification of human uses and users of ecosystems and their end-products.

It is important to note is that while there are important parallels between NAICS/NAPCS and NESCS, there are important differences as well. Specifically, it is important to note that while NAICS and NAPCS provide alternative ways for classifying economic goods and services,

NESCS-S and NESCS-D together constitute the classification system for FFES. They are complementary systems that need to be used in conjunction with each other to identify and classify FFES.

Section 4.3 provides a detailed description of the NESCS structure, including the various subcomponents of NESCS-S and NESCS-D. Section 4.4 summarizes key elements of NESCS.

4.2   Conceptual Framework for the NESCS

In this section, we describe how the basic tools and concepts from national economic accounting can be adapted and applied to develop a classification system for ecosystem services. Even though the current focus of NESCS is not an application to green GDP accounting, for several reasons the NIPA framework remains particularly useful for developing an ecosystem services classification system. First, developing NIPAs requires an understanding of the interconnected system of input-output relationships that make up the economy. The flow and transformation of products and services through this network ultimately results in goods and services that are consumed by "households." This interconnected system can be expanded to include the input-output relationships that (1) make up natural ecosystems and (2) connect ecosystems to humans. Second, although the NAICS and NAPCS economic classification systems were primarily created to support NIPA's development, we argue that many of the concepts, principles, and lessons learned from the development of NAICS/NAPCS are transferable to NESCS. More broadly, similar data sources and information may be relevant for cost-benefit analysis.

Developing the conceptual framework for NESCS involves three broad steps, which are described in the following sections. First (Section 4.2.1), we describe the conceptual framework underlying the economic classification systems and the input-output relationships. Second (Section 4.2.2), we describe how this framework can be expanded to classify ecosystem services. Given that our primary objective is to support marginal analysis, the next step (Section 4.2.3) is to adapt the framework so that it can be applied for conducting such analyses. This motivates and provides the theoretical underpinning for the NESCS structure (described in Section 4.3). It is important to note here that a more complete representation of the utility function should also include other elements that contribute to welfare such as societal services (e.g., family relations, health, etc.). However, since this is not the focus of NESCS, we do not explicitly include such elements in our framework in the interest of simplicity.

4.2.1   The Conceptual Framework for Economic Goods and Services

Before discussing their implications for ecosystem services classification, in this section we describe and illustrate the conceptual framework underlying the national economic accounts and classifications systems. To illustrate the key concepts relevant for NESCS, we develop a simple, scaled-down representation of the framework underlying the national product accounts (shown in Figure 4-1). The process begins with the main inputs—the factors of production—which are represented here by physical capital and human labor. These two factors are represented as **stocks**. That is, they can be measured at a specific point in time rather than over a period of time. For example, the capital stock represents the amount of machinery and equipment available to be used in production on a given date, and the labor stock represents the number of workers available.

Figure 4-1.   Conceptual Framework for Classification of Economic Goods and Services

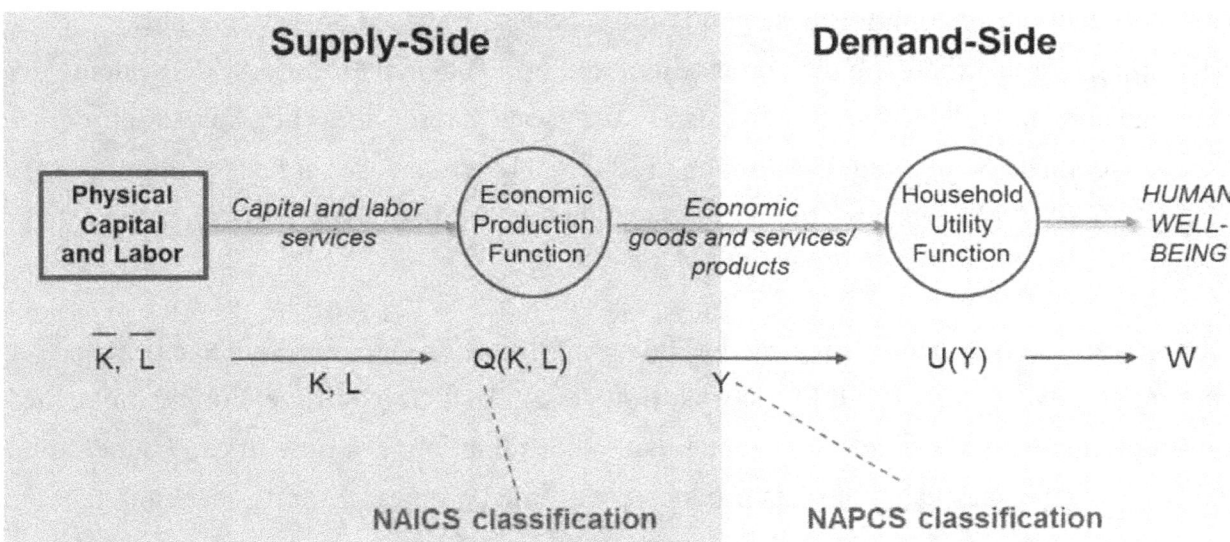

In this conceptual model, the productive factors are represented as inputs to a **production function**, Q(K, L). This function represents the various processes and technologies by which labor and capital inputs are transformed into output products (Y, represented by Y = Q(K, L)). As a simplification, this representation of the production process does not include other inputs such as services from land, energy, and other natural resources; however, these elements will be included later in this section.[49]

---

[49]   In essence, this simplified representation of the production process only shows the "value added" by labor and capital inputs. The contribution of other inputs are not shown but are essential to the output Y.

These output products—goods and services—are sold to consumers who use them as consumption inputs to support their own well-being. The process by which the consumption inputs (Y) are transformed into human well-being is represented by a **utility function**, U(Y). Because well-being (W, represented by W = U(Y)) is an abstract concept that cannot be measured, it is sometimes approximated by consumption, in other words, it is in some cases assumed that W = Y. Again, it is important to note that we assume that utility is only a function of the consumption inputs (Y) here for the sake of simplicity.

In this simplified framework, the *supply-side* of the economy is represented by the production process on the left hand side of Figure 4-1, which results in the output of goods and services (Y). The *demand-side* is represented by the consumption and utility generation process on the right side of the figure, which begins with the same goods and services (Y), which are now consumption inputs. As discussed in more detail below, the NAICS system is designed to classify the production function for these goods and services (Y) based on a supply-side perspective (i.e., who is producing the commodities and how?), whereas the NAPCS system focuses on the demand-side perspective to classify the goods and services (i.e., how and by whom are the products being used?). Although production functions/uses are not explicitly defined in NAICS/NAPCS systems, these criteria are embedded in how the categories are developed.

In Figure 4-1, stocks are represented as boxes and functions are represented as circles. **Flows** are represented as arrows. Unlike stocks, flows are measured over time. For example, the output from production (Y), which is also the input to consumption, is a flow. It can be measured as the number or monetary value of units produced/consumed per year. Strictly speaking, the arrows from the capital and labor stock box to the production function represent flows of **labor and capital services**. Although it is relatively easy to measure labor and capital stocks at a given time, it is inherently more difficult, particularly for capital, to measure the amount of services being provided to the production process in a particular period. Therefore, for practical reasons, the stock measures $(\overline{K}, \overline{L})$ of these inputs (e.g., number of employed workers and value of installed machinery) are often used as proxies for the service flows in the production function.

One important expansion of this simple conceptual framework is to distinguish between intermediate and final products. **Intermediate goods and services** are the outputs produced by one sector of the economy, which are then used as production inputs in another sector. For example, many agricultural commodities such as corn are sold and used as inputs for food processing and other industries. In contrast, **final goods and services** are sold directly to consumers (i.e., households) and not used to produce other goods and services for the market economy.

In Figure 4-2, this expansion of the conceptual framework is represented by splitting the production process into two production functions. The production of intermediate products is represented by $Y_I = Q_I(K, L)$, and the production of final good and services is represented by $Y_F = Q_F(K, L, Y_I)$. Only the final products, Y_F, are treated as consumption goods and services and are therefore inputs to the utility function $W = U(Y_F)$. As in Figure 4-1, for simplification the first stage of production (in this case, "intermediate" production) only includes capital and labor as inputs. Other inputs not produced by humans, such as from land and other natural resources, are not yet included in the framework.[50]

Figure 4-2 emphasizes that some producers act as both **demanders** of goods and services (Y_I) and as **suppliers** of goods and services (Y_I and Y_F). The production and sale of final products is represented in the figure as an area of overlap between the supply and demand sides.

Figure 4-2. Conceptual Model Distinguishing Between Intermediate and Final Goods and Services Production

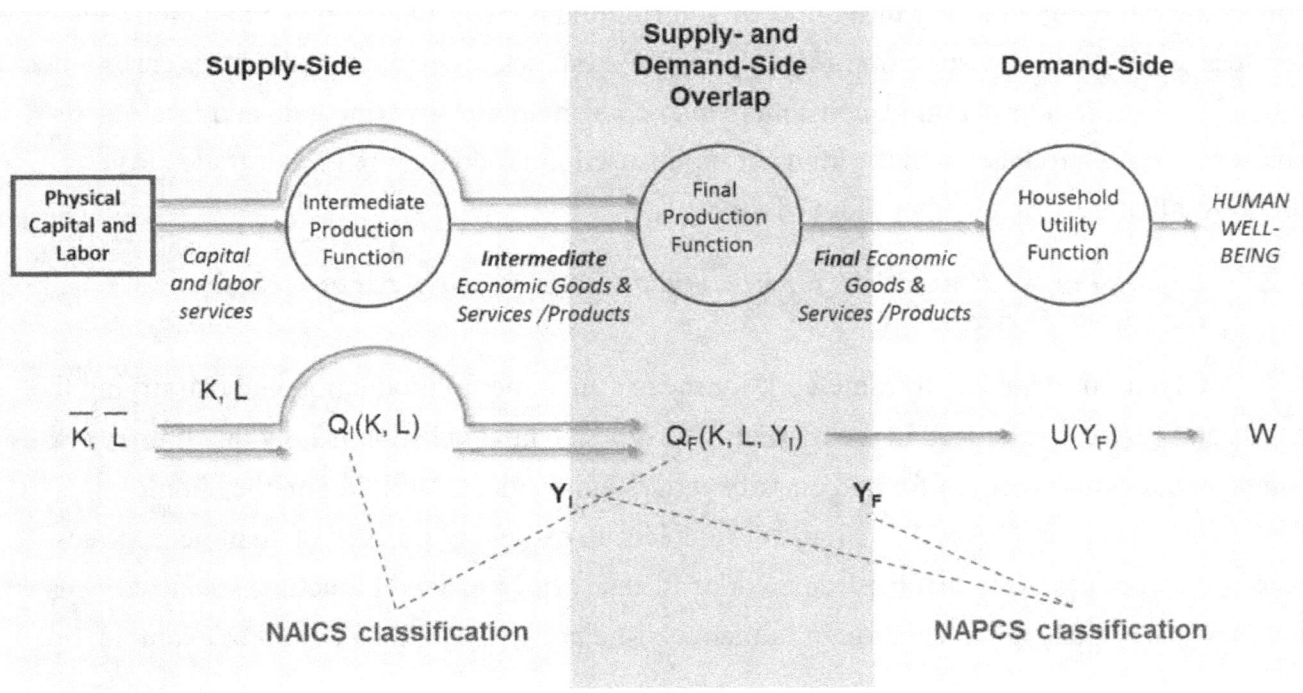

---

[50] By including intermediate production outputs (Y_I) as inputs in final production, the representation of final production now includes more than just the "value added" input contribution of labor and capital.

Although not shown explicitly in Figure 4-2, some types of goods and services can be classified as both intermediate and final. This dual classification occurs because assigning goods and services as final or intermediate depends on who demands them, not on who supplies them. For example, corn sold to food processors is classified as an intermediate good, whereas corn sold to consumers is a final good.

The distinction between intermediate and final products is essential for estimating the value of total production in an economy. Most importantly, the possibility of double counting the value of a product (e.g., once when a commodity is sold as an intermediate product and again when it is sold as part of a final product) is avoided if the analysis only focuses on the value of final products. In other words, the value of a final product includes the value of *all* intermediate goods and services that have contributed to its production (as well as the value-added contributions of capital and labor to intermediate and final production). Adding together the value of intermediate and final products would be redundant.

This framework also assumes that only final goods and services contribute *directly* to human welfare (through the household utility function). Intermediate goods and services only contribute *indirectly* to utility through their contribution to the production of final goods and services.[51] We use this assumption for simplification but acknowledge that there are cases where it may not be true. For example, consumers may care about how intermediate and final goods and services are produced and the kinds of inputs used in production (e.g., fair trade coffee, dolphin safe tuna, organically produced food).

4.2.2 Expanding the Framework for Economic Goods and Services to Include Ecosystem Services

Given this conceptual framework for measuring national production and classifying the goods and services produced in the market economy, the next question is how this framework can be expanded to account for and classify ecosystem services. Building on the simple framework shown in Figure 4-2, Figure 4-3 depicts an expanded framework that incorporates ecosystem services. Note that the framework can easily be expanded to include social services for a more complete representation of human well-being. This could be done, for example, by

---

[51] Using the terminology from Herrendorf et al. (2013), this representation of utility can be described as a "final expenditure" approach. The alternative—a "value-added" approach—assumes that individuals care about at least some of the intermediate inputs that go into the production of a final good or service. Both approaches have merit and are consistent with the generally accepted proposition that households derive value from goods and services through the "bundles" of attributes they provide (Lancaster, 1966). However, as should become clear in the next section of this report, the final expenditure approach has advantages for expanding the framework to include ecosystem services.

including a circle for Social Services Production that would be connected to the household utility function by an arrow. However, since social services are not the focus of NESCS, we do not include this in the NESCS framework.

The expansion includes three main elements. First, in addition to physical capital and labor, it includes a stock of **natural capital**, N, representing all ecosystems. Second, it includes an "**ecological production function**," $Q_E(N)$, which represents the myriad of natural processes through which ecosystems (N) transform, adapt, and evolve to produce ecological end-products (E, represented as $E = Q_E(N)$). Boyd (2007) describes these end-products as biophysical features of ecosystems that are (1) concrete, tangible, and measurable, and (2) of direct value to humans. These end-products are conceptually similar to stocks of FEGS (see Table 2-6, here) in Landers and Nahlik (2013).

Figure 4-3. Expanded Conceptual Framework, Including Ecological Production and Flows of Final Ecosystem Services (FFES) as Inputs to the Economy

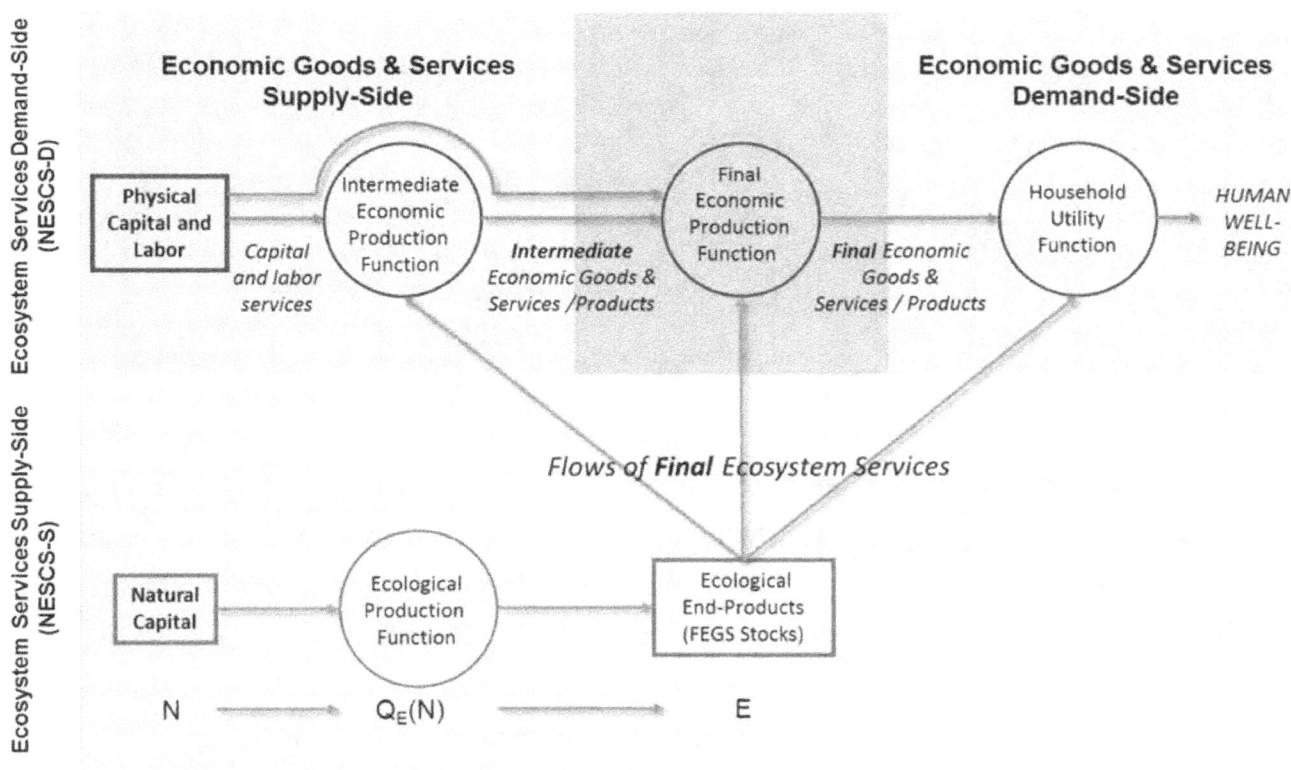

For example (illustrated in Figure 4-4), wetlands can be thought of as natural capital assets. Among the many natural processes supported by wetlands, they receive and filter surface water flows and recharge groundwater aquifers. Through this multistage ecological "production" process, they replenish and maintain stocks of clean, accessible water. These water resources represent important ecological end-products for direct human use. For example, water is used to support different types of plant cultivation such as corn. Water can also be used for industrial processing activities such as manufacturing of cornflakes, for example. Water can also support human life and health in different ways, such as being drawn from an aquifer for drinking.

Figure 4-4. Example Illustrating Conceptual Framework

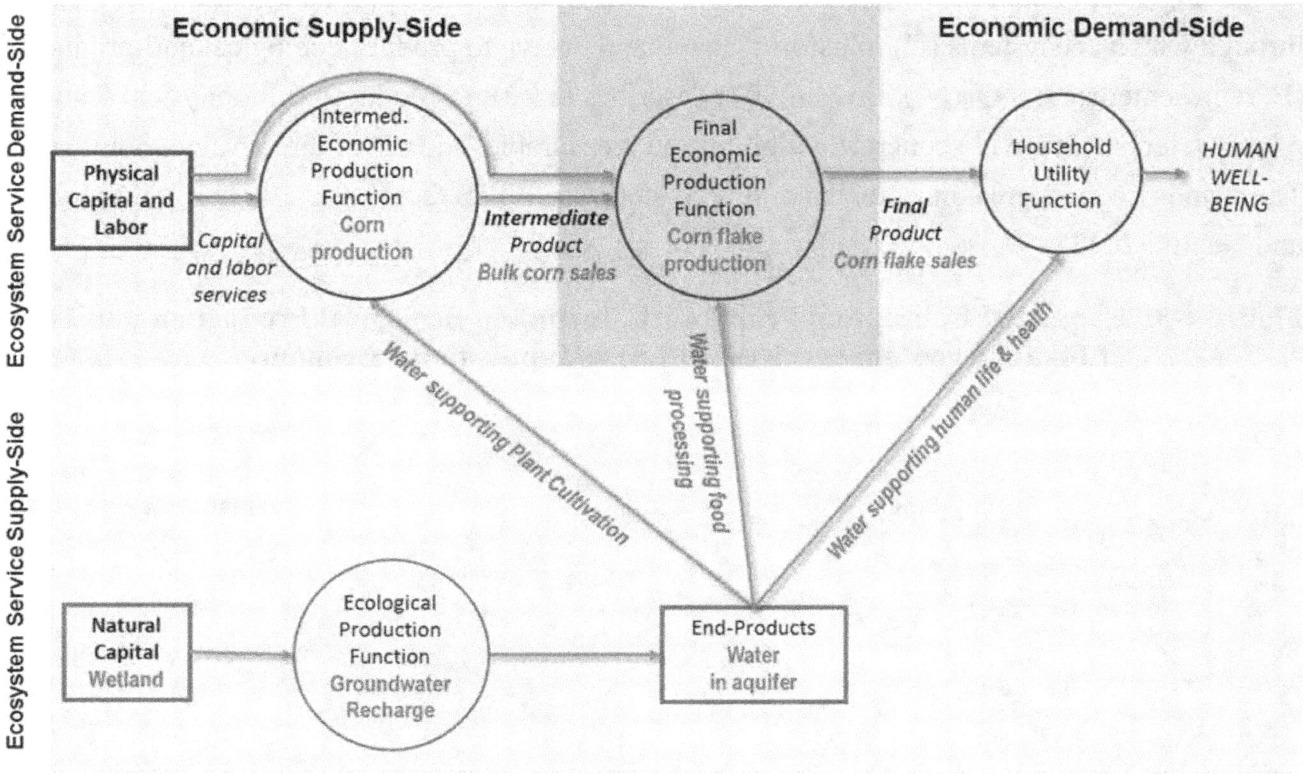

Similar to the economic production function, $Q(K, L)$, the ecological production function, $Q_E(N)$, can involve multiple stages of **intermediate ecosystem services** production. These intermediate production stages are not shown in Figure 4-4; however, using a similar approach to what is shown in Figure 4-2, they could be separated from final production (See Figure B-1 in Appendix B). For example, the process of surface water filtration by wetlands can be thought of as an ecological production process that provides an intermediate ecosystem service. The removal of contaminants from water flowing into aquifers is clearly important for humans; however, this service is not directly used by humans. Instead, it is the services provided by the aquifer, as a source of clean drinking water supply, that are of direct value to humans and that therefore constitute final services.

Box 4-1. NESCS Definitions

End-products are biophysical components of nature that are either directly used by humans to produce goods and services or directly enjoyed or used to yield human well-being. They can usually (but not always) be interpreted as stocks of ecological goods.

> Example: Stocks of clean water in an aquifer

Flows of Final Ecosystem Services (FFES) are the contributions of nature (1) directly to human production processes or (2) directly to households and human well-being. FFES occur at the point of hand-off between natural systems (ecosystems) and human systems (producers and households). They are represented as service flows between ecological end-products and direct human uses. Note that by definition, ecosystem services only exist when they contribute to human well-being.

> Example: Water directly extracted from freshwater sources to support plant cultivation, food processing, and human health/well-being (as drinking water)

Intermediate *ecosystem* **services** are inputs to the natural processes that ultimately produce FFES.

> Example: Wetlands' removal of contaminants from water flowing into aquifers

Intermediate *economic* **goods and services** are produced using human inputs (physical capital and labor) and ecological inputs (FFES) and are sold to other producers. They are the outputs produced by one sector of the economy, which are then used as production inputs in another sector.

> Example: Agricultural crops used as inputs in food processing such as corn used to produce ethanol

Final *economic* **goods and services** are produced using human inputs (physical capital and labor), intermediate economic goods and services (e.g., corn) and ecological inputs (FFES) and are sold to households who use them as consumption inputs to support their own well-being. They are not used to produce other goods and services for the market economy.

> Example: Food products sold to consumers, such as cornflakes

---

*NOTE: Flows of final ecosystem **goods** are not included or defined in the NESCS framework. The main reason for this exclusion is that the process of transferring physical ecosystem products from nature to humans, which is necessary to generate flows of goods, typically requires human inputs. For example, transferring portions of existing timber or fish stocks to humans for their use requires human labor for harvesting. In our framework, the involvement of human inputs implies that the transferred goods are classified as economic rather than ecosystem goods.*

Third, it includes **flows of final ecosystem services** (FFES) from the ecological end-products to the human systems. Importantly, all human systems—producers and households—are only shown on the demand-side of ecosystem services. Although humans produce and supply economic goods and services, they do not supply ecosystem services. The production and supply of end-products is strictly the domain of natural systems. Through policies and other actions, humans can alter the *functioning* of these natural systems, but ultimately it is the *natural* systems that supply the end-products. Thus, considering human effects on natural systems is not essential for *defining and classifying* flows of ecosystem services from natural *to* human systems.

Although there are important parallels between final *economic* goods and services (Y) and final *ecosystem* services, a few key differences should be noted. First, final economic goods and services are defined at the point of exchange between producers (firms) and households. A good or service is final when it is not intended to be used as an input in further production and resold. In contrast, final ecosystem services are defined at the point of "hand-off" between ecological systems and human systems. They are final if no further additional biophysical transformation is required for humans to see them as relevant to human well-being (Ringold et al., 2009). This final hand-off is shown in Figure 4-4 in the arrows that extend from the ecological end-products to the human production and utility functions.

Second, whereas a final economic exchange can be directly observed through the terms of a market transaction (i.e., the number and price of units exchanged), no explicit transaction exists in a transfer from an ecosystem to humans. Therefore, the determination of a final ecosystem service is more open to interpretation than it is for an economic good or service.

As shown in Figure 4-3, end-products enter human systems in two main ways: as inputs to market production activities; or as direct inputs to households[52] (non-market sector) and human well-being.

In the first case, this role is shown by (1) the arrow from the ecological end-products (E) to the market production function, and (2) the addition of FFES as an input to production, such that $Y = Q(K, L, FFES)$. For example, returning to the wetland and aquifer example shown in Figure 4-4, if a private drinking water supplier uses the aquifer as a water source, then it is the direct recipient of final ecosystem services from the stock of available groundwater (the ecological end-product, in this case). Using labor, capital, and the water available in the aquifer, the supplier then withdraws water and produces a final economic service (Y = tap water distribution, measured in gallons per day) to a local population.

In the second case, the role as input to the household/non-market sector is shown by (1) the arrow from the ecological end-products (E) to the utility function, and by (2) the addition of FFES as an argument in this function, such that $W = U(Y, FFES)$. In this case, the example might be a household with a private well connected to the aquifer. Rather than purchasing tap water distribution services from a supplier, this household is the direct recipient of the final ecosystem services offered by the groundwater resource.

Given these distinctions, it is important to note that *flows of final ecosystem **goods*** are not included or defined in this framework. The main reason for this exclusion is that the process of transferring physical ecosystem products from nature to humans, which is necessary to generate flows of goods, typically requires human inputs. For example, transferring portions of existing timber or fish stocks to humans for their use requires human labor for harvesting. In our

---

[52] Potentially, end-products may also provide inputs to other sectors such as government.

framework, the involvement of human inputs implies that the transferred goods are classified as economic rather than ecosystem goods.

Similar to the services received from physical capital, it can be difficult to specifically define and measure the flow of final services received from an ecological end-product. In these cases, the ecological end-product, like the stock of physical capital, can be a reasonable proxy.[53] In other words, the production function using the final ecosystem service as an input can be expressed as $Y = Q(\overline{K}, \overline{L}, E)$ and, similarly, the utility function can be expressed as $W = U(Y, E)$. For example, consider the aesthetic amenities provided to local residents by a clear-water lake. The lake is the ecological end-product and the residents are the direct users, but it is difficult to quantify the flow of final ecosystem services received each year. In this case, a stock measure of the ecological end-product (e.g., number of lake acres) may be the best available indicator.

Figure 4-3 also divides the overall system into a supply and demand side; however, in this expanded framework, the green side represents the natural assets and processes that supply FFES, and the blue side represents the human processes that use and derive well-being from the FFES. As represented in Figure 4-3, the categorization of the supply-side providers of FFES in this context is referred to as NESCS-S. The demand side categorization of uses of FFES is referred to as NESCS-D.

It is important to emphasize and acknowledge that the strict separation of natural (green) and human (blue) systems greatly oversimplifies the actual relationship between humans and their natural surroundings. However, this dividing line is included in the conceptual framework because it provides a useful abstraction for representing the basic properties of FFES. It stresses that, to identify *final* ecosystem services, one must consider where the relevant "point of hand-off" occurs between natural and human systems.

In actuality, there are numerous ways in which human and natural systems overlap and interact. In many ways, humans can be viewed as an integral part of the larger ecosystem, and there are few remaining ecosystems with little or no human footprint. Examples of "gray" areas where it can be difficult to separate the two systems are:

- urban ecosystems, such as parks and greenways, which are often actively managed or even created by humans;

---

[53] It is important to note here that both measures of quantity as well as quality of end-products matter since both these attributes are important factors that contribute to flows of services.

- agricultural and commercial forest ecosystems, which produce economic goods and services (e.g., crops, timber) but also generate positive non-market externalities such as aesthetically pleasing scenery and wildlife habitat;

- wetland and stream restoration projects, which are constructed by humans with the express purpose of restoring and providing lost ecosystem services;

- national parks, forests, and recreation areas, where ecosystems are typically less managed than the previous examples, but they still involve significant human input; and

- ocean fisheries, which are increasingly depleted or otherwise modified by commercial and recreational fishing.

These gray areas present unavoidable challenges for identifying FFES. Section 6 of this report discusses these types of challenges (and some potential solutions) in more detail. Some of these gray areas can also be thought of as examples of "feedback" flows from human to natural systems, which are discussed in the next subsection and in Appendix B.

4.2.3 A "Marginal" Analysis Framework for Applying NESCS

The previously described framework describes the linkages between ecosystems and human welfare and defines the role of ecosystem services. The next step is to adapt the framework so that it can be used to conduct "marginal" analyses of environmental policies, like cost-benefit analysis (CBA).[54] In marginal analysis, the focus is on how policy-related changes to ecosystems affect human well-being. From the perspective of CBA, this primarily means expanding the framework to identify, and to the extent feasible, quantify and monetize the benefits of policies that protect or enhance ecosystems. For other analyses, like cost-effectiveness analysis (CEA), the focus is on quantifying the positive effects of policy changes (i.e., benefits), but not necessarily expressing them in monetary terms.

At its most simple level, this adaptation means using the framework to focus on changes in its main components and how these changes are linked. For changes that occur through market production systems, the linked changes can be represented as:

---

[54] NESCS can also support analysis of policies outside the environmental context that also result in changes to ecosystems.

$$Policy\ Action \rightarrow \Delta N \rightarrow \Delta E \xrightarrow[\Delta FFES]{} \Delta Y \rightarrow \Delta W.$$

The policy action is assumed to first cause a change in the natural capital stock (N), such as an increase in the number of wetland acres.[55] This change ultimately leads to a change in one or more ecological end-products (E), such as an increase in the amount of water stored in an aquifer that is used for irrigation and an increase in surface water storage capacity, which reduces the number of flooding events. Another example of a change in E might be a change in contamination levels in the water that is extracted from rivers for food manufacturing purposes. Changes to these end-products result in an increase in flows of services from the end-products, and consequently in the final market goods and services (Y). For instance, an increase in groundwater stocks can support higher rates of irrigation and higher food production, and decreases in flooding can result in a larger supply of housing. An increase in the quality of water can result in a larger supply of drinking water and other beverages. These additional economic goods and services then result in a higher level of human utility/well-being (W). In practice, this linked process can be more complicated if, for example, changes in the costs of production and prices are incorporated, but the fundamental connections remain.

For ecosystem changes that affect households directly rather than through the market production system, the linked changes can be represented as:

$$Policy\ Action \rightarrow \Delta N \rightarrow \Delta E \xrightarrow[\Delta FFES]{} \Delta W$$

In these cases, household utility is directly enhanced by the change in ecological end-products, for example by reducing the risk of flood damage to homes.

In a CBA framework, the change in well-being (ΔW) is typically converted to monetary terms by estimating households' maximum willingness to pay (WTP) for either the additional market final goods and services (ΔY) or the additional non-market final ecosystem service flows (ΔFFES). A more formal and mathematical representation of the link between a policy action that impacts on one or more ecosystems and the resulting effects on human well-being is provided in Appendix A.

One of the main reasons for developing NESCS is to help identify and group all of the *multiple pathways* through which changes in ecosystems (ΔN) result in changes in human well-

---

[55] In practice, policy actions could alter the ecological production process or directly change the ecological end-product, but for simplicity and illustrative purposes, we show the impact through ΔN.

being (ΔW). Identifying these pathways is a necessary first step for developing a comprehensive accounting of benefits.

Several types of pathways are shown in Figure 4-5. Each involves two or more stages. In the first stage, the figure shows that a specific change in one ecosystem (ΔN) can affect multiple ecological end-products, as shown by the arrows to ΔE_1, ΔE_2, ..., ΔE_n. For example, a change in wetland acres can lead to changes in groundwater supplies, changes in flood risks, changes in migratory waterfowl populations, and to changes in the stock of clean water.

In the following stage(s), each of these affected end-products can then affect utility via changes in FFES in one or more ways. First, the end-products can affect utility directly, as shown by the arrows to ΔW_{j+3}, ..., ΔW_{j+p}. For example, the changes to groundwater, flooding, and lakes can each have a separate but direct effects on households' well-being. Second, the end-products can affect utility indirectly, through their effects on the market production processes, as shown by the arrows to ΔY_1, ΔY_2, ..., ΔY_m. For example, changes in groundwater and surface water storage can affect the production of food crops and public water supply systems.

Figure 4-5. Representation of Multiple Pathways Linking Policy-Related Ecosystem Impacts (ΔN) to Changes in Human Well-Being (ΔW)

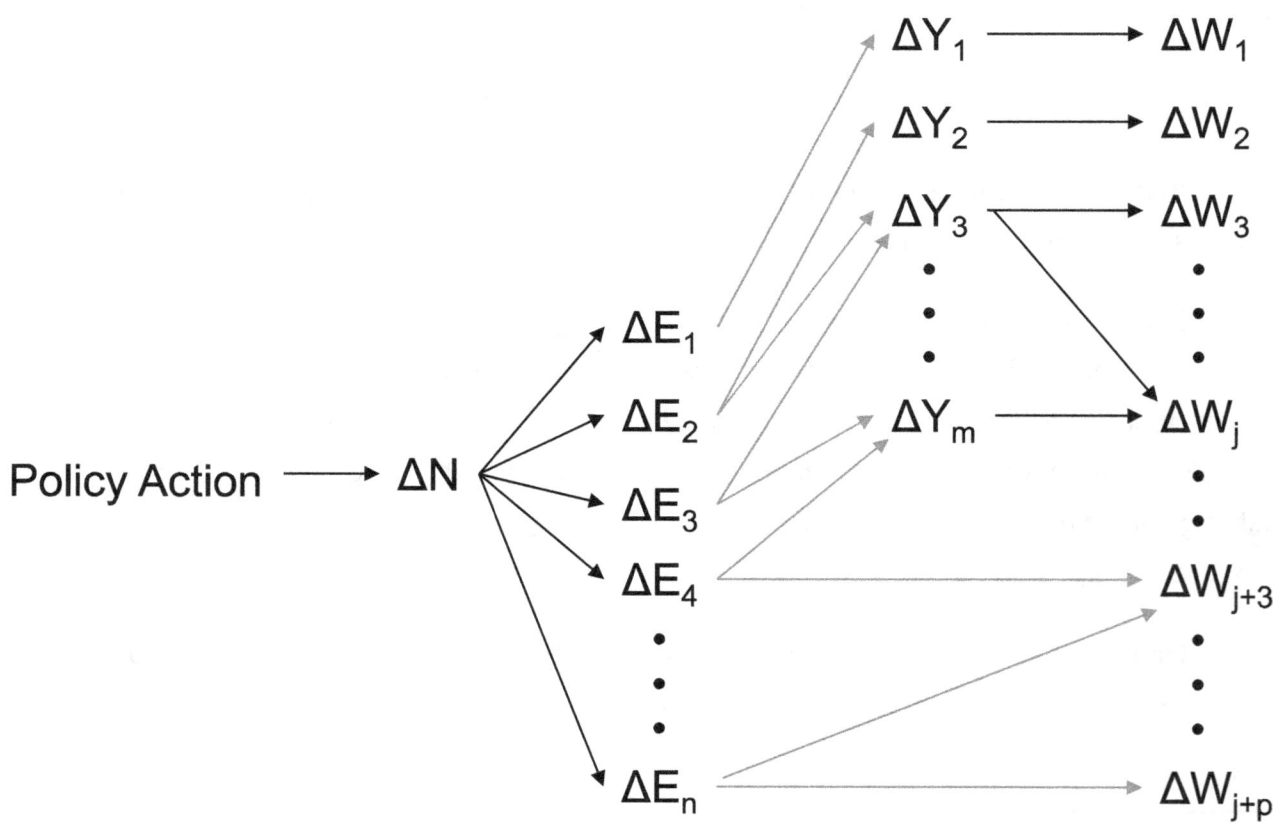

The number of distinct pathways between ΔN and ΔW can be expansive because (1) multiple ecological end-products can be affected, (2) each end-product can have multiple pathways to utility (i.e., each end-product can yield multiple FFES through direct utility changes to households and indirect utility changes through market production systems), and (3) each market production process can be affected by multiple end-products. The NESCS system should provide a classification structure for identifying and grouping distinct pathways that are potentially relevant. Developing this system requires answers to the following questions:

1. Which ecosystems "produce" which ecological end-products?

2. What ecological end-products are potentially relevant for each of the different market production sectors and directly relevant for households?

3. How are end-products directly used to produce market goods and services and/or to directly derive utility?

4. Who are the direct users of these end-products?

Once the potentially relevant pathways are identified, the next step in applying the NESCS framework in a CBA is to quantify the input-output relationships that make up these pathways. In other words, it requires quantitative models to (1) represent the affected ecological production functions, market production functions, and utility (i.e., valuation) functions, and to (2) quantify the magnitude of the input-output relationships. However, these steps are beyond the scope of NESCS.

When moving to the next steps of quantifying and valuing the relevant input-output relationships identified with NESCS, it is important to note that the *multiple* links shown in Figure 4-5 do not necessarily represent the *complete* set of links and effects on human well-being from a policy action. In many instances, the connections will be more complex and multidimensional for at least two reasons. First, within the human economy there may be "spillover" or "general equilibrium" effects between economic sectors (including households). For example, a change in the production process for one type of economic good (Y_i) can alter input and output prices in the market in a way that results in ripple effects through other sectors of the economy. A specific example would be a policy that increases instream flows to hydro-electric facilities, which would increase electricity supplies and put downward pressure on prices in the electricity market. These price changes would not only affect the users of hydropower, but could also affect other (e.g., fossil fuel) producers of electricity and their customers. For illustration, these linkages could be represented by additional arrows between the ΔYs shown in Figure 4-5 (as well as arrow to sectors not originally shown in the figure).

Second, there may be "feedback" effects from the affected human systems back into the natural systems. For example, a policy that increases the fish stocks in an estuary would increase the FFES to commercial fishermen that harvest from the estuary. However, these changes could also result in a significantly higher number of vessels fishing in the estuary, which would increase catch levels and offset some or all of the increased fish stock. For illustration, these linkages could be represented by additional arrows from the ΔYs (and ΔWs) back to the natural capital.

Although this added complexity can be critical for fully capturing the magnitude and value of the ecosystem service changes caused by a policy change, it is also important to emphasize that this added complexity does not require revisions or additions to the NESCS structure. First, capturing spillover effects within the human economy does not require a different classification of FFES inputs. Second, although feedback effects may return to the human systems through different pathways than the original primary effects, there is no reason to believe that these different pathways require an alternative classification structure for FFES.

4.3   Proposed Classification Structure and Coding System for NESCS

In this section, we describe the main components of the NESCS structure, which can be used to identify the pathways between ecosystems and human welfare. We also describe the NESCS coding system that allows for a numeric representation of the NESCS structure.

As described in previous sections, we use and adapt many of the broad principles underlying NIPA, NAICS, and NAPCS to define the NESCS structure. First, as described in Section 4.1, similar to NAICS and NAPCS,[56] NESCS also distinguishes between a supply-side and a demand-side grouping.

Second, NIPA, NAICS, and NAPCS provides a framework and tools to account for the main input/output relationships in the economy. For example, these relationships and the links between the NAICS and NAPCS categories are represented in tables (referred to in NIPA as the "make" and "use" tables) for the economy. Similar tables can be developed to show the relationship between NESCS-S and NESCS-D.

---

[56]   Note again that while NAICS and NAPCS are two separate systems, NESCS-S and NESCS-D are components of the same system.

Third, NAICS/NAPCS uses a coding or numbering system that represents the underlying classification structure and allows for easy identification and referencing of unique industries/commodities. Both of these classifications allow for a nested hierarchical structure, where each hierarchical level represents an aggregation of the components of the lower level. Currently, NAICS industries can be represented by two-digit codes at the most aggregate level and six-digit codes at the most disaggregate level. NAPCS commodities can be represented by three-digit codes at the most aggregate level, and seven–digit codes at the most disaggregate level. These coding systems were designed to allow for flexibility in several ways. They allow policy analysts to select the level of aggregation (or digit code) that is appropriate for their needs. They are also flexible enough to allow for additional categories to be included at each hierarchical level, and for additional detailed levels to be included as the need arises. Similar logic can be applied to NESCS, to develop a numbering system that will allow for easy identification and referencing of each unique pathway.

We draw from the NIPA and NAICS/NAPCS structures and coding systems and adapt them to define a structure and a coding system for NESCS-S/NESCS-D, as summarized in Table 4-1. To address the requirements for marginal analysis identified in Section 4.2.3, we define four main classification groups. The first two groups—the environment and end-product groups—are contained within NESCS-S, and the last two groups—the direct uses/non-use and direct user groups—are within NESCS-D.

Within each of these four groups, NESCS adopts a nested hierarchical structure so that each group can be represented at multiple levels of aggregation or detail. In the current NESCS structure, as many as three hierarchical levels are defined within each group—Class, Subclass, and Detail. Across these four groups, each hierarchical structure is independently defined from the others. The classification structure representing the four groups (with examples of classes and subclasses), and the flows between them, is represented in Figure 4-6. Below, we describe this classification structure in more detail.

Similar to NAICS/NAPCS, our goal is to adopt the general principles of a classification system (described in Section 4.3.1), so that at each level of the hierarchy, all subgroupings represent mutually exclusive categories. The main purpose of this feature is to avoid double counting, particularly when the system is used for CBA or other types of ecosystem service accounting.

Table 4-1. NESCS Structure and Coding System

	NESCS–S		NESCS–D	
Group	**Environment**	**End-Product**	**Direct Use/Non-Use**	**Direct User**
Definition	Spatial units with similar biophysical characteristics that are located on or near the Earth's surface, and that contain or produce "end-products"	Biophysical components of nature that are directly used or appreciated by humans	Different ways in which end-products are used or appreciated by humans	Entities that directly use or appreciate the end-products
Hierarchy and Coding System NESCS Code for FFES*: WW. .YYYY.ZZZ/ZZZ				
Class	W	WW.X	WW.XX.Y	WW.XX.YYYY.Z
Subclass	WW	WW.XX	WW.XX.YY	WW.XX.YYYY.ZZZ
Detail			WW.XX.YYYY	WW.XX.YYYY.ZZZZZZZ
Example 1: Water in the ocean being used as a medium for freight transportation NESCS Code for FFES: 15.12.1202.1483111				
Class	Aquatic: **1**	Water: **1**	Direct Use: **1**	Industry: **1**
Subclass	Open Ocean and Seas: **15**	Liquid Water: **12**	In-Situ Use: **12**	Transportation and Warehousing: **148**
Detail			Transportation medium: **1202**	Deep Sea Freight Transportation: **1483111**
Example 2: Water in rivers being extracted for household gardening purposes NESCS Code for FFES: 11.12.1105.201				
Class	Aquatic: **1**	Water: 1	Direct Use: **1**	Households: **2**
Subclass	Rivers and Streams: **11**	Liquid Water: **12**	Extractive Use: **11**	Households: **201**
Detail			Support of plant or animal cultivation: **1105**	

ᵃ Note that this 15-digit code is the most disaggregated level of representation. Different levels of aggregation can be used depending on the context (See Examples 1 and 2 for different levels of aggregation for users).

NESCS also defines a coding system for this classification structure by assigning digits to represent categories for each group. To represent that the categories across the four main groups are independent of one another, NESCS uses a decimal point between the digits of each of the four groups. The highest level of aggregation (Class) is represented by a single digit. The levels below (Subclass and Detail) are represented by additional digits.

This classification and coding structure is designed to provide flexibility for expanding the system in the future. As necessary, it can be expanded to include (1) additional categories at each existing aggregation level, and (2) additional levels of detail for each existing category. The NESCS structure also provides the option of using different levels of aggregation for each group, depending on the context. For example, one group can be represented at the highest level of aggregation (Class), two groups can be represented at the second level of aggregation (Subclass) and the fourth group can be represented at the most disaggregated level (Detail). Table 4-1 shows the maximum possible disaggregation level and corresponding digits for each group.

The purpose of this classification structure is to define unique FFES categories, such that each category represents a distinct pathway for linking changes in ecosystems to changes in human welfare. Using this NESCS structure, a unique FFES category is defined by combining elements from each of the four groups. In other words, each FFES category is composed of an environmental class, an ecological end-product class, a direct use/non-use class, and a direct user class. At its most disaggregate level for all four groups, an FFES can therefore be identified and referenced by a unique fifteen-digit code.

4.3.1 Proposed Structure for NESCS-S

Since FFES is defined as a flow from a producer to a consumer, it is important to first identify natural systems that are "producers" of the service and this is done in NESCS-S. Thus, the system follows the broad supply-side logic of the economic classification systems. This leads to the classification structure of the first group—Environment. This group includes two possible levels of aggregation—Environmental Class and Subclass. The categories and coding system for each of the two levels are obtained from the FEGS-CS (Landers and Nahlik, 2013). The different Environmental Classes/Subclasses are defined as spatial units, with similar biophysical characteristics, that are located on or near the Earth's surface and can be interpreted as producers of end-products. The subclasses can be viewed as spatial units with similar biophysical characteristics (see section 3.2 of Landers and Nahlik, 2013 for details). The FEGS-CS structure for this group is reproduced in Table 4-2.

In developing the NESCS-S classification, we also reviewed and considered other existing classifications for the environmental classes and subclasses shown in Table 4-2 (e.g., wetlands, rivers and streams, near coastal marine). In most cases, we identified numerous classification systems built for a variety of purposes and using different organizing criteria (a summary is available on request). For example, some systems apply purely biophysical criteria and others apply human use-based criteria for defining categories and subcategories. For this stage of NESCS-S development, we concluded that it would not be productive to define and include more detailed levels than Environmental Classes and Subclasses.

NESCS-S also includes a classification structure for the second group—End-Products. Following Boyd and Banzhaf (2007), **end-products** are biophysical outcomes of nature that humans directly use and care about. This is a key component of the NESCS structure, since it identifies the point of hand-off between ecosystems and human systems. Thus, this component helps distinguish between "final" and "intermediate."

Figure 4-6. NESCS 4-Group Structure

Figure 4-6. NESCS 4-Group Structure

Environment

Aquatic
- Rivers and streams
- Wetlands
- Lakes and ponds
- Near coastal marine
- Open ocean and seas
- Groundwater

Terrestrial
- Forests
- Agroecosystems
- Created greenspace
- Grasslands
- Scrubland/shrubland
- Barren/rock and sand
- Tundra
- Ice and snow

Atmospheric
- Atmosphere

End-Products

Water
- Snow/ice
- Liquid water

Flora
- Specific classes/species of flora

Fauna
- Specific classes/species of fauna

Other Biotic Components
- Specific types of natural material

Atmospheric Components
- Air
- Solar light/radiation

Soil
- Specific types of soil

Other Abiotic Components
- Specific types of natural material

Composite End-Products
- Scapes: views, sounds and scents of land, sea, sky
- Regulation of extreme events
- Presence of environmental class

Other End-Products

Stock Indicators, Flow Indicators, Quality Indicators, Site Indicators, Indicators Characterizing Extreme Events

NESCS-S

Flows of Final Ecosystem Services

Direct Use/Non-Use

Use
- **Extractive Use**
  - Raw material for transformation
  - Fuel/energy
  - Industrial processing
  - Distribution to other users
  - Support of plant or animal cultivation
  - Support of human health and life or subsistence
  - Recreation/tourism
  - Cultural/spiritual activities
  - Information, science, education, and research
  - Other extractive use
- **In-situ Use**
  - Energy
  - Transportation medium
  - Support of plant or animal cultivation
  - Waste disposal/assimilation
  - Protection or support of human health and life
  - Protection of human property
  - Recreation/tourism
  - Cultural/spiritual activities
  - Aesthetic appreciation
  - Information, science, education, and research
  - Other in-situ use

Non-Use
- **Existence**
- **Bequest**
- **Other Non-Use**

Direct User

Industries
- Agriculture, Forestry, Fishing and Hunting
- Mining
- Utilities
- Construction
- Manufacturing
- Wholesale Trade
- Retail Trade
- Transportation and Warehousing
- Information
- Finance and Insurance
- Real Estate Rental and Leasing
- Professional, Scientific, and Technical Services
- Management of Companies and Enterprises
- Administrative Support and Waste Management and Remediation Services
- Educational Services
- Health Care and Social Assistance
- Arts, Entertainment, and Recreation
- Accommodation and Food Services
- Other Services

Households

Government

NESCS-D

84

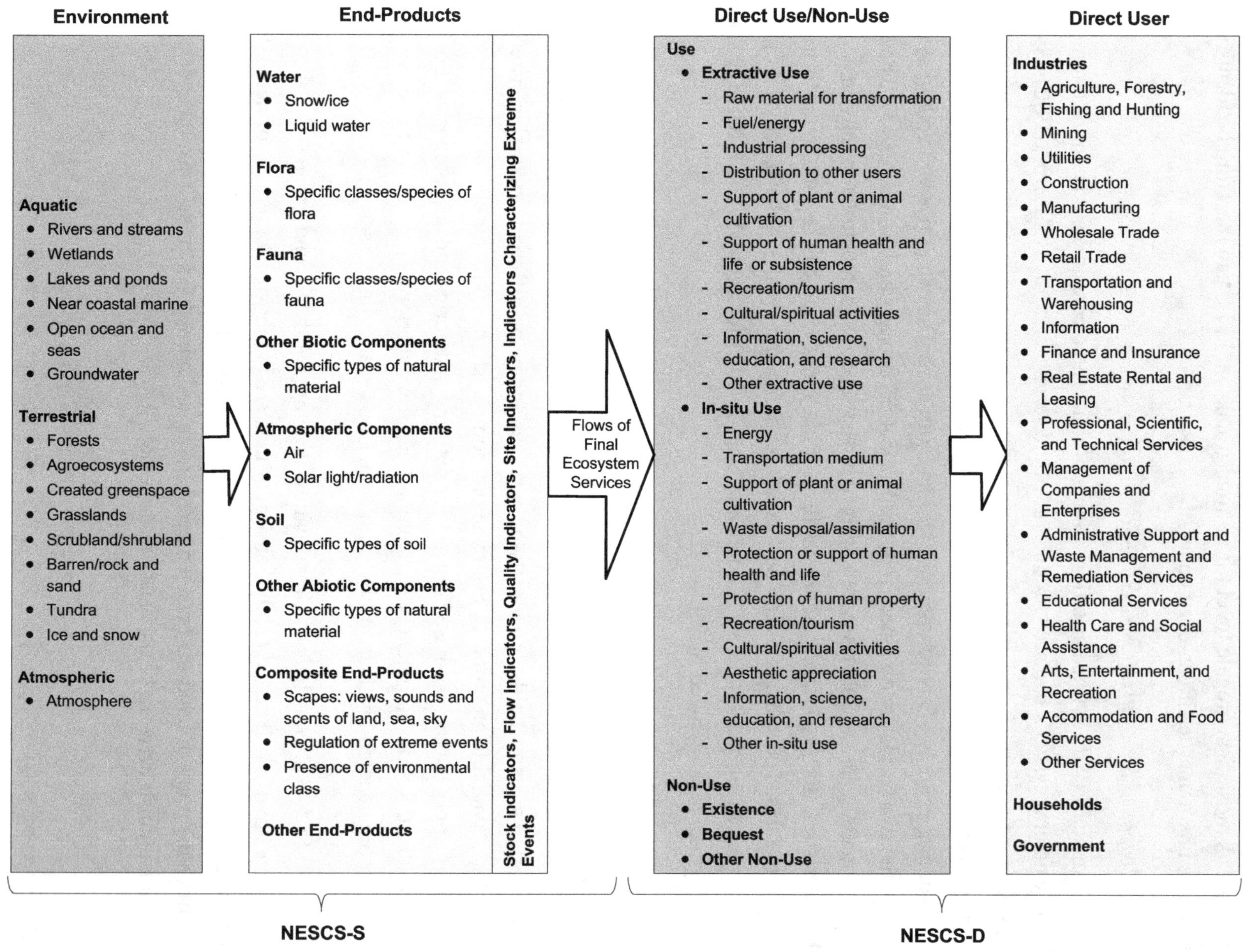

Table 4-2. Classification of Environment[a]

Environmental Class	Environmental Subclass
1. Aquatic	11. Rivers and Streams
	12. Wetlands
	13. Lakes and Ponds
	14. Near Coastal Marine
	15. Open Ocean and Seas
	16. Groundwater
2. Terrestrial	21. Forests
	22. Agroecosystems
	23. Created Greenspace
	24. Grasslands
	25. Scrubland / Shrubland
	26. Barren / Rock and Sand
	27. Tundra
	28. Ice and Snow
3. Atmospheric	31. Atmospheric

[a] The environmental classes and subclasses were obtained from Landers and Nahlik (2013).

It is important to note that what is "final" depends on the context. In different contexts, the same component of nature can be either intermediate or final (like many economic goods and services). For example, water is an end-product when we consider drinking water, but, for recreational fishing uses where fish is the relevant end-product, water can be considered to be an intermediate product that is essential for fish abundance. Thus, what is "final" is specific to the ways they are used by human beings. These uses are the domain of NESCS–D, which is described in the next section.

Similar to Environment, this classification group also includes two potential aggregation levels—End-Product Class and Subclass. Landers and Nahlik (2013) identify 21 "FEGS Categories" (reproduced in Table 2-6). We use this FEGS-CS list as our starting point for identifying end-products. For this classification group, one of our objectives for NESCS was to add more structure to the list provided in FEGS-CS, so that it identifies mutually exclusive categories to the extent feasible. The NESCS End-Product Classes are defined in Table 4-3, which also includes descriptors and examples for the End-Product Subclasses.

There are nine possible End-Product Classes in NESCS.[57] While NESCS does include End-Product Subclasses as the next hierarchical level, it is important to note that not all of these

---

[57] To allow for flexibility to include other types of end-products in the future (that NESCS currently does not include), we include a ninth category "Other end-products." This category does not include any Subclasses, descriptors, or examples at this point

Classes are further decomposed into Subclasses. Specifically, Flora, Fauna, Other Biotic Components, Soil, and Other Abiotic Components are not decomposed further. We only provide descriptions and examples of these Classes. This is because, similar to Environmental Subclasses, End-Product classes can be decomposed in a myriad of different ways. For example, the Subclasses of Flora and Fauna could potentially be represented as individual species or as classes of species, based on the objective of the user. Flora could be decomposed into Subclasses based on life form,[58] growth habit,[59] ethnobotany (e.g., Ghimire and Aumeeruddy-Thomas, 2009), ecological succession, water requirements, or other factors. Similarly, Fauna could be grouped by taxonomy or habitat (e.g., Lindenmayer and Cunningham, 1996). Soils can also be classified in a variety of ways. For example, classification based on soil texture recognizes the distinction between particles sizes (i.e., clay, sand, silt, gravel, pebbles).[60] Soil taxonomy groups soils based on similar physical and chemical properties.[61] For this stage of NESCS-S development, we concluded that it would not be productive to define and include End-Product Subclasses other than those shown in Table 4-3.

Defining categories for End-Product Classes and Subclasses that are mutually exclusive is challenging since there can be substantial complexity and diversity in which elements or aspects of nature that people care about. There are two primary reasons for this complexity:

1. People may care about individual end-products, but they may also care about combinations of them. There are complementarities in consumption where the value of the bundle is higher than the sum of the individual parts. For example, people may value an entire landscape more than the flora, fauna, water, and other parts. Also, different people may care about different combinations of different end-products.

2. Some people may care about specific attributes of end-products, rather than the end-products themselves, as we have defined them. For example, a person who enjoys fall color viewing may care about the vibrant colors of the foliage of a tree rather than about the full tree itself.

To account for the first issue, we include a category called "Composite end-products." This category reflects the fact that specific combinations of end-products (within or across Environmental Classes), rather than isolated end-products by themselves, are valuable to human

---

[58] http://www.fs.fed.us/database/feis/plants/ (accessed May 29, 2015).
[59] http://plants.usda.gov/about_adv_search.html (accessed May 29, 2015).
[60] ftp://ftp.wcc.nrcs.usda.gov/wntsc/H&H/training/soilsOther/soil-USDA-textural-class.pdf (accessed May 29, 2015).
[61] http://www.nrcs.usda.gov/wps/portal/nrcs/main/soils/survey/class/ (accessed May 29, 2015).

beings. The composite end-products class, represented by "8," includes three Subclasses. The first Subclass (represented by "81") includes different types of "Scapes" (landscapes, seascapes, skyscapes) and can potentially encompass multiple Environmental Classes. This category captures the fact that human beings may value entire landscapes more than the individual components comprising them, such as trees, birds, and wildlife. "Regulation of extreme events" (represented by "82") is considered a Composite End-Product because it encompasses multiple Environmental classes and End-Product Classes. For example, the ability of a system to regulate a flood is influenced by the soil and the flora. The fire susceptibility of a system is influenced by the atmosphere, vegetation, water, and soil. Landslides are regulated by soil, water, and vegetation. The third Subclass (represented by "83") is "Presence of Environmental Class," and accounts for the fact that a person may care about the presence of an entire Environmental Class (e.g., tropical forests), rather than individual end-products of the class (e.g., trees, birds, etc.).

Table 4-3. Classification of End-Products[a]

End-Product Class	Definition	End-Product Subclass	End-Product Subclass Examples
1. Water	Liquid and solid forms of water	11. Snow/Ice	
		12. Liquid water	
2. Flora	All plant, fungal and unicellular life	Specific classes/species of flora	• trees • shrubs • herbs • grasses • ferns • mosses • lichens • mushrooms • viruses • bacteria
3. Fauna	All animal life	Specific classes/species of fauna	• mammals • fish • birds • reptiles • amphibians • insects
4. Other Biotic Components	All other biota or biotic material that are not part of or attached to a currently living floral or faunal source	Specific types of natural material	• driftwood not attached to currently living tree • shells not attached to currently living clams
5. Atmospheric Components	Components of the atmosphere (excluding categories described above)	51. Air	• oxygen • carbon dioxide • helium • nitrogen • hydrogen

(continued)

Table 4-3. Classification of End-Products[a] (continued)

End-Product Class	Definition	End-Product Subclass	End-Product Subclass Examples
		52. Solar light/radiation	
6. Soil	The unconsolidated mineral or organic matter on the surface of the Earth	Specific types of soil	• mud • clay • loam • stones • rocks
7. Other Abiotic Components	Other abiotic material (cannot be attributed to soil, atmosphere or water)	Specific types of natural material	
8. Composite End-Products	A composite set of specific elements and components of single or multiple environmental classes	81. Scapes: • views • sounds and scents of land, sea, sky or a combination	• seascape • landscape • skyscape • includes natural phenomenon (e.g., geysers, hot springs, sunsets, cloud formations) and subterranean features , etc.
		82. Regulation of extreme events	Regulation of: • floods • fire • landslides • storms
		83. Presence of environmental class/subclass	Presence of tropical forests
9. Other End-Products	All other end-products (nec*)		

[a] For some end-products, we do not develop categories for subclass. We only include a descriptor and a few examples. (See text for details)
* Not elsewhere classified

When applying NESCS to identify unique pathways for valuation, one important note of caution must be kept in mind when considering Composite End-Products. Because they represent a combination of elements, many of which may also be thought of as individual and separate End-Products, particular care must be taken to avoid double counting their ecosystem service values. For example, a natural landscape, which can be thought of as a Composite-End-Product that delivers ecosystem services to households (Direct Users) through aesthetic appreciation (Direct Use), may also be comprised of individual elements such as wildlife (fauna), trees (flora), and lakes (water). The ecosystem services provided by this landscape would be double counted if they were included as both (1) a flow from the Composite End-Product (landscape) to households through aesthetic appreciation, and as (2) individual flows from the separate End-Products (wildlife, trees, lakes) to the same households through aesthetic appreciation. Therefore, the analyst must make a judgment about whether to treat the landscape as a single

Composite End-Product or as multiple individual End-Products for the Direct Use-User group in question (but not both without taking extra care to avoid double-counting, even if this involves subtracting a redundant identified FFES in an actual valuation application).[62] In this case, NESCS offers a flexible framework for defining end-products, but care must be taken to use it appropriately to avoid double counting.

To address the second issue (individuals caring about attributes of the end-products rather than the entire end-product), we argue that in further developing and applying NESCS, it will be important to define indicators that characterize different attributes of end-products. The attributes of end-products that people care about will depend on how the end-product is being used, as well as who is using it. Different types of indicators for these attributes would include the following:

- stock indicators;

- flow indicators;

- quality indicators;

- site characteristic indicators; and

- indicators to characterize extreme events such as floods, etc.

Table 4-4 illustrates the detailed structure of NESCS-S for each of the three Environmental Classes (Aquatic, Terrestrial, and Atmospheric). A few key end-products are identified for each of these classes, and the End-Product Subclass category includes examples drawn from the FEGS-CS.

Table 4-5 provides additional detail for NESCS-S by identifying some key end-products in each Environmental Subclass. This table can be viewed as a type of "Make Table" where the Environmental Class "produces" the end-product. Figure 4-7 shows how NESCS-S can also be represented with the end-product categories nested below the Environmental classes, in a "tree structure."

---

[62] Note that double counting occurs in these cases because the Composite and individual End-Products are enjoyed or used by the same Direct Use-Direct User group (i.e., aesthetic appreciation by households). If the Direct Use or Direct User categories are different across the Composite and individual End-Product categories (e.g., landscapes are enjoyed by households for aesthetic appreciation while wildlife is used by the same households for recreational hunting) then double counting is less likely to be an issue. However, it is again ultimately the user of NESCS who will have to make the judgment about how to apply the End-Product categories to avoid duplication of ecosystem service values.

Table 4-4. NESCS-S Detailed Structure: Examples

Environmental Class	Environmental Subclass	End-Product Class	End-Product Subclass Examples
1. Aquatic	11. Rivers and Streams	1. Water	
		2. Flora	• grasses • reeds • wild rice • watercress • water pepper
		3. Fauna	• fish • crawfish • clams • snails • alligator • beaver • moose • ducks • geese
		4. Other Biotic Components	• fish oil • shells • mollusk shells
		6. Soil	
		7. Other Abiotic Components	
		8. Composite End-Products	• birds singing • rapids • viewscapes • presence of the environmental class

(continued)

90

Table 4-4. NESCS–S Detailed Structure: Examples (continued)

Environmental Class	Environmental Subclass	End-Product Class	End-Product Subclass Examples
2. Terrestrial	24. Grasslands	2. Flora	• berries • tubers, grasses • flowers, seeds • fungi
		3. Fauna	• ducks • rabbit • deer • elk • buffalo • bison • grasshoppers • fox • wolf • coyotes • different species of pollinators, depredators and (pest) predators
		4. Other Biotic Components	• deer antler velvet • eggs • dried flowers
		6. Soil	
		7. Other Abiotic Components	
		8. Composite End-Products	• viewscapes • sounds and scents • presence of the environmental class
3. Atmospheric	31. Atmospheric	5. Atmospheric Components	• wind • weather
		1. Water	
		3. Fauna	• birds
		8. Composite End-Products	• thunder • wind blowing • clouds • sunsets • viewscapes • presence of the environmental class

Table 4-5. End-Products in Each Environmental Class

End-Product Class	End-Product Subclass	11. Rivers and Streams	12. Wetlands	13. Lakes and Ponds	14. Near Coastal Marine	15. Open Ocean and Seas	16. Groundwater	21. Forests	22. Agroecosystems	23. Created Greenspace	24. Grasslands	25. Scrubland / Shrubland	26. Barren / Rock and Sand	27. Tundra	28. Ice and Snow	31. Atmospheric
1. Water	11. Snow/Ice	X	X	X	X	X									X	
	12. Liquid water	X	X	X	X	X	X									X
2. Flora	Specific classes/species of flora	X	X	X	X	X		X	X	X	X	X	X	X	X	
3. Fauna	Specific classes/species of fauna	X	X	X	X	X		X	X	X	X	X	X	X	X	X
4. Other Biotic Components	Specific types of natural material	X	X	X	X	X		X	X	X	X	X	X	X		
5. Atmospheric Components	51. Air															X
	52. Solar light/radiation															X
6. Soil	Specific types of soil	X	X	X	X	X		X	X		X	X	X	X		
7. Other Abiotic Components	Specific types of natural material	X	X	X	X			X		X	X	X	X	X	X	
8. Composite End-Products	81. -Scapes: • views	X	X	X	X	X	X	X	X	X	X	X	X	X	X	X
	• sounds and scents of land, sea, sky or a combination															
	82. Regulation of extreme events	X	X	X	X	X	X	X	X	X	X	X	X	X	X	X
	83. Presence of environmental class/subclass	X	X	X	X	X	X	X	X	X	X	X	X	X	X	X
9. Other End-Products																X

Figure 4-7. NESCS-S Tree Structure

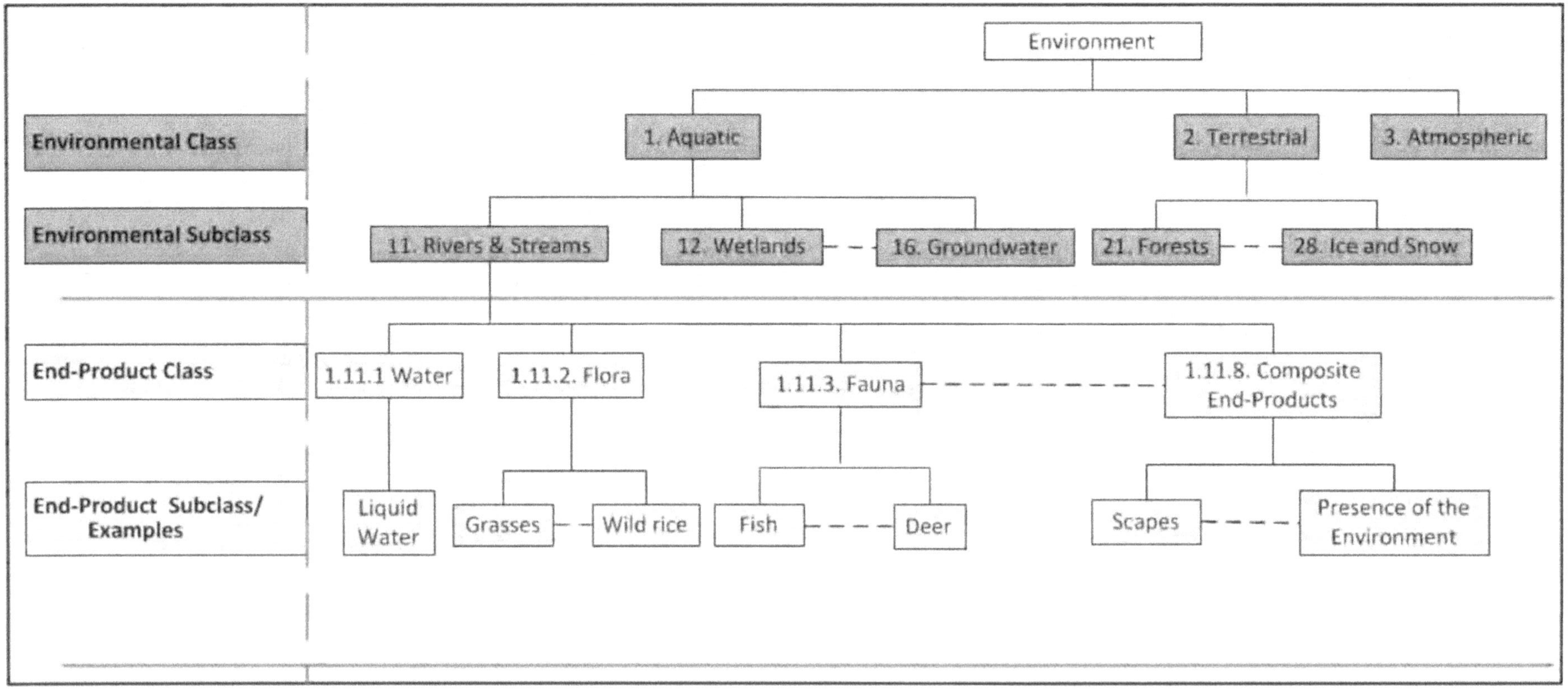

In summary, NESCS-S helps to identify the point of hand-off from ecosystems to human beings, by defining specific Environmental Classes and the End-Product Classes provided by them. However, these ecosystem supply-side components must be combined with human demand-side components to fully identify the potential pathways between ecosystems and human well-being. The demand-side orientation (NESCS-D) is the focus of the next subsection.

4.3.2   Proposed Structure for NESCS-D

NESCS-D supports the second step in identifying direct contributions of ecosystems to human welfare, such that it will then support quantification and valuation of changes in ecosystem services. NESCS-D helps identify ways in which end-products are used by human beings. In this section we propose a structure for NESCS-D classification. This classification follows the broad demand-side logic of the NAPCS commodity classification. NESCS-D defines explicitly how FFES are directly used and by whom they are used/appreciated.

One way in which the NESCS-D differs from the NAPCS classification system is that, whereas NAPCS includes categories for intermediate economic goods and services (represented by $Q_I(K,L)$ in Figure 4-2), NESCS-D does not include intermediate ecological processes and services (represented by the function $Q_E(N)$ in Figure 4-2). For example, the products consumed by business are inherently intermediate economic inputs; however ecosystem services primarily consumed by businesses are inherently final ecological inputs to production. Intermediate ecological services, such as nutrient cycling, which is important but not directly used by humans, are deliberately not included in NESCS-D.

The first part of NESCS-D (third group in the overall NESCS structure) is a classification of Direct Use/Non-Use, which defines different ways in which End-Products are directly used or appreciated by humans. This group has three hierarchical levels—Class, Subclass and Detail. The Direct Use/Non-Use categories are defined in a way that is broadly consistent with the total economic valuation (TEV) framework (shown in Figure 4-8), which is a commonly used organizing and conceptual framework for non-market valuation. Use–Non-Use Classes distinguishes between Use (represented by "1") and Non-Use (represented by "2"). Direct Use/Non-Use Subclasses distinguish between extractive (represented by "11") and in-situ uses (represented by "12"), and existence (represented by "21"), bequest (represented by "22"), and other non-uses (represented by "23").[63]

---

[63]   The TEV framework is also sometimes expanded to include a separate "option value" category; however, there is a growing consensus in the environmental economics literature that option value arises out of uncertainty

Use/Non-Use Detail further decomposes the Subclasses into categories that strive to be mutually exclusive and exhaustive. Table 4-6 lists and defines the different hierarchical levels in this group. In developing these use/non-use categories it is important to distinguish between how end-products are used and what they are used to produce. For example, irrigation is a direct use and crop production relates more to the NAICS direct user category. Some direct use categories may apply to multiple different end-products (e.g., the end-products water and air are both directly used for energy).

The second part of NESCS-D (fourth NESCS group) is a classification of Direct Users. The distinction between use and user classifications is included to account for the fact that some direct use categories may apply to multiple different direct user categories (e.g., direct use of water for industrial processing could apply to many NAICS categories). Moreover, some direct users of ecosystems may benefit from multiple uses. This use-user dichotomy is similar, for example, to established classification structures such as those adopted by the U.S. Census Bureau (Table 3-5) and the United Nations. As can be seen from this table, the same user could be linked to different uses (column) and the same use could be linked to different users (row).

We adopt the terms "Direct Use/Non-Use" and "Direct User" to distinguish between users who *directly* use or appreciate end-products from *potential* downstream users. For example, commercial fishermen who extract fish and then sell to households (represented by the Direct Use "Distribution to other users") are Direct Users of the end-product fish. Households who purchase fish from them are downstream users. Including commercial fishers and households who buy fish would entail double counting. Hence, we do not include downstream uses and uses in the NESCS.

Direct user categories representing sectors that directly use (or have non-use values for) ecological end-products thus comprise the fourth group in the NESCS structure. This group also has three hierarchical levels—Class, Subclass, and Detail. Again, we follow established classification structures similar to those adopted by the U.S. Census Bureau and the United Nations. The first hierarchical level or "Class" thus includes the broad sectors of the economy— Industry, Household, and Government (represented by digits "1," "2," and "3," respectively).

---

regarding future supply or demand of the commodity in question, and it requires an expected utility approach for incorporating uncertainty. Consequently, it should not be interpreted as a separate category of value, nor is its introduction essential for classifying ecosystem services.

Figure 4-8. Valuation Framework (TEV)

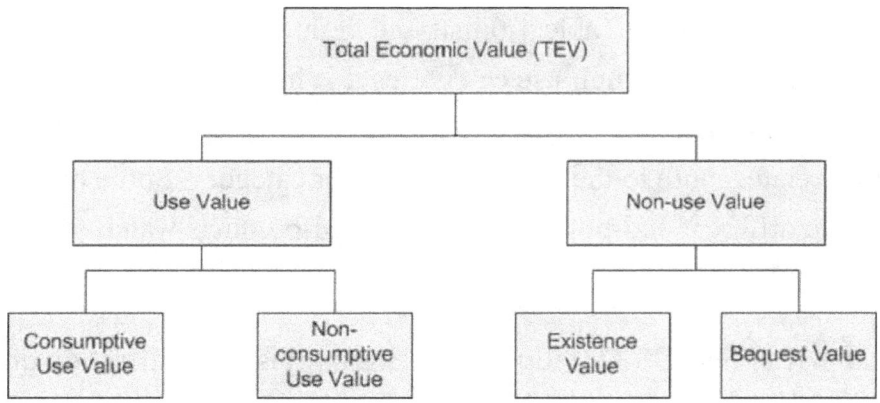

Table 4-6.   Classification of Direct Use/Non-Use

Direct Use/Non-Use Class	Direct Use/Non-Use Subclass	Direct Use/Non-Use Detail	Direct Use/Non-Use Detail Definition
1. Direct Use	11. Extractive Use	1101. Raw material for transformation	Extracted or harvested and transformed into other commercial products
		1102. Fuel/energy	Extracted or harvested and directly used as an energy source for commercial production
		1103. Industrial processing	Extracted or harvested and directly used in other ways as a material in industrial processing
		1104. Distribution to other users	Extracted or harvested for distribution to other users
		1105. Support of plant or animal cultivation	Extracted or harvested to support human cultivation of plant or animal life
		1106. Support of human health and life or subsistence	Extracted or harvested and directly used by humans for subsistence, health, or other life support
		1107. Recreation/tourism	Extracted or harvested as part of an outdoor recreational or nature tourist activity
		1108. Cultural/spiritual activities	Extracted or harvested as part of a non-recreational cultural or spiritual activity
		1109. Information, science, education, and research	Extracted or harvested to directly support scientific research or education
		1199. Other extractive use	Extracted or harvested for other uses
	12. In-situ Use	1201. Energy	Used in situ as a source of energy for commercial production
		1202. Transportation medium	Used in situ as a medium for transporting goods or humans
		1203. Support of plant or animal cultivation	Used in situ to support human cultivation of plant or animal life
		1204. Waste disposal/assimilation	Used in situ as a sink for assimilating and disposing of waste
		1205. Protection or support of human health and life	Used in situ to protect against damages or otherwise support human health and life
		1206. Protection of human property	Used in situ to protect against damages to human property
		1207. Recreation/tourism	Used in situ as part of an outdoor recreational or nature tourist activity

(continued)

Table 4-6. Classification of Direct Use/ Non-Use (continued)

Direct Use/Non-Use Class	Direct Use/Non-Use Subclass	Direct Use/Non-Use Detail	Direct Use/Non-Use Detail Definition
		1208. Cultural/spiritual activities	Used in situ as part of a non-recreational cultural or spiritual activity
		1209. Aesthetic appreciation	Used in situ for aesthetic (visual and other senses) appreciation, separate from outdoor/nature recreational, tourist, cultural or spiritual activities
		1210. Information, science, education, and research	Used in situ to directly support scientific research or education
		1299. Other in-situ use	Used in situ for other purposes
2. Non-Use	21. Existence	2101. Existence	Appreciated and valued by humans for existence reasons (without direct use or contact)
	22. Bequest	2201. Bequest	Appreciated and valued by humans for bequest reasons (without direct use or contact)
	29. Other non-use	2901. Other non-use	Appreciated and valued by humans for other reasons (without direct use or contact)

Table 4-7. Classification of Direct Users

Direct User Class*	Direct User Subclass
1. Industry	111. Agriculture, Forestry, Fishing and Hunting
	121. Mining
	122. Utilities
	123. Construction
	131–33. Manufacturing
	142. Wholesale Trade
	144–45. Retail Trade
	148–49. Transportation and Warehousing
	151. Information
	152. Finance and Insurance
	153. Real Estate Rental and Leasing
	154. Professional, Scientific, and Technical Services
	155. Management of Companies and Enterprises
	156. Administrative and Support and Waste Management and Remediation Services
	161. Educational Services
	162. Health Care and Social Assistance
	171. Arts, Entertainment, and Recreation
	172. Accommodation and Food Services
	181. Other Services (except Public Administration)
2. Households	201. Households
3. Government	301. Government

* Last two digits for Industry (in the Subclass Column) represent 2-digit NAICS sectors. We omit NAICS 814 (Private Households) and NAICS 92 (Public Administration). See https://www.census.gov/cgi-bin/sssd/naics/naicsrch?chart=2012 for definitions. We include separate categories for households and government to ensure we capture a broader range of uses than that implied by the NAICS definition.

The NAICS system, which provides an existing and well-established framework for classifying production establishments into mutually exclusive categories, is used to further decompose market-based users of ecosystem services into Subclasses. Thus, the coding system for User Subclass for Industry is represented by three digits, where the first digit is "1," and the next two digits reproduce the two-digit NAICS structure. [64] For example, the NAICS code for Agriculture, Forestry, Fishing and Hunting is "11" and the code for Mining is "21." The User Subclass codes for these two sectors are "111" and "121," respectively.

The next hierarchical level, User Detail, further decomposes Industry into more detailed categories. This level is represented by "1," followed by four or more NAICS digits. Since NAICS can include up to six possible digits, User Detail can include up to seven digits, depending on the level of disaggregation necessary. For example, "Deep Sea Freight Transportation" has a NAICS code of 483111 and thus its counterpart in NESCS User Detail is represented by "1483111." If a less detailed category was needed or applicable, the NESCS counterpart for NAICS sector "Deep Sea, Coastal, and Great Lakes Water" is represented by "14831." If an even more aggregate sector is needed, the NESCS counterpart for the NAICS sector "Water Transportation" is represented by "1483" Transportation. Thus, the structure of the User Detail allows for flexibility in the level of NAICS sector aggregation. Table 4-7 lists the categories for User Class and Subclass. This table does not reproduce the NAICS categories for User Detail (in the interest of space), since this is already provided by the U.S. Census Bureau. [65]

Currently, NESCS does not decompose Households and Government any further. Similar to Table 3-5, all households are grouped into a single user category. The main reason for this single grouping of households is because it is inherently more difficult to separate households into mutually exclusive categories. In contrast, market establishments tend to specialize in the production of specific goods or services; therefore, they can be relatively easily separated into mutually exclusive categories, based on their "primary activity." For example, the primary activity of a hotel is lodging, but it may also include a restaurant whose primary activity is serving meals and beverages. In the automotive industry, dealers maintain sales and service facilities in the same location. The NAICS grouping of establishments according to specialization in similar production practices is carried over to NESCS-D, grouping them as similar users of ecosystem services. We emphasize that the single category for households does not imply that the benefits directly derived by households are any less important or smaller than those experienced by businesses. Table 4-8 shows how the I-O use table framework in Table 3-5

---

[64] See http://www.census.gov/eos/www/naics/faqs/faqs.html#q5 for details.
[65] Available at https://www.census.gov/cgi-bin/sssd/naics/naicsrch?chart=2012 (accessed May 29, 2015).

can be adapted to show the relationship between the *uses* and the *users* of FFES. Rather than using the table rows to show the NAPCS commodity categories, it instead includes the preliminary NESCS-D classification. However, the classification of users across the table columns is the same as in Table 3-5.

Similar to Table 3-5, Table 4-8 can also be used to present a cross-walk between uses and users. For example, specific uses, such as cooling water provided by rivers and flood protection provided by wetlands, can be relevant for multiple user categories. Similarly, specific user groups may have multiple uses for the same ecological end-product.[66]

Table 4-8 provides a framework to identify and, if necessary, store the FFES values for different user groups.

4.3.3 Relationship between NESCS-S and NESCS-D: Incorporating NESCS Into an Input-Output Framework

Similar to the NAICS/NAPCS classification, the NESCS-D demand-side classification can be linked to a NESCS-S supply-side classification. Table 4-9 shows how the framework in Table 3-4 can be adapted to show the relationship between demand- and supply-side classifications for ecosystem services. In Table 4-9, the NAICS production sectors are replaced by NESCS-S categories of ecosystems and ecological end-products. The NAPCS commodity classifications are replaced with the NESCS-D classification shown.

As a starting point, this table can be used to illustrate the **cross-walk** between the NESCS-D and NESCS-S categories. For each NESCS-S ecological end-product category (column), the table can indicate the NESCS-D direct uses (rows) it supports. For example, the freshwater fauna (e.g., fish) category can link to distribution to other users (e.g., commercial fishing), raw material for transformation (e.g., for food supply purposes), recreation (recreational fishing), or to non-use values. Similarly, for NESCS-D direct use categories (rows), the table can be used to indicate the ecological end-products that support it. For example, the non-use category can be linked to (i.e., supported by) all of the ecological end-product categories shown.

---

[66] At the lowest and most detailed levels of the NESCS-D hierarchy, it is likely that the number of user categories associated each use category will approach one. At this level of disaggregation, therefore, the distinction between use and user will be less relevant and important.

Table 4-8. An Example of a NESCS Table Relating Use/Non-Use and Users

Direct Use/Non-Use Class	Direct Use/Non-Use Subclass	Direct Use/Non-Use Detail	111. Agriculture, Forestry, Fishing and Hunting	121. Mining	122. Utilities	123. Construction	131-33. Manufacturing	142. Wholesale Trade	144-45. Retail Trade	148-49. Transportation and Warehousing	151. Information	152. Finance and Insurance	Total market sector use	201. Households	301. Government	Total non-market sector use
1. Direct Use	11. Extractive Use	1101. Raw material for transformation	✓	✓		✓	✓	✓						✓		
		1102. Fuel/energy			✓	✓	✓	✓						✓		
		1103. Industrial processing			✓	✓	✓		✓							
		1104. Distribution to other users	✓													
		1105. Support of plant or animal cultivation	✓													
		1106. Support of human health and life or subsistence												✓		
		1107. Recreation/tourism	✓				✓		✓					✓		
		1108. Cultural/spiritual activities														
		1109. Information, science, education, and research														
		1199. Other extractive use														

(continued)

100

Table 4-8.   An Example of a NESCS Table Relating Use/Non-Use and Users (continued)

Direct Use/Non-Use Class	Direct Use/Non-Use Subclass	Direct Use/Non-Use Detail	111. Agriculture, Forestry, Fishing and Hunting	121. Mining	122. Utilities	123. Construction	131-33. Manufacturing	142. Wholesale Trade	144-45. Retail Trade	148-49. Transportation and Warehousing	151. Information	152. Finance and Insurance	Total market sector use	201. Households	301. Government	Total non-market sector use
1. Direct Use (cont.)	12. In-situ Use	1201. Energy			√									√		
		1202. Transportation medium	√						√					√		
		1203. Support of plant or animal cultivation	√	√	√	√	√	√	√	√	√	√		√		
		1204. Waste disposal/assimilation		√								√		√		
		1205. Protection or support of human health and life							√					√		
		1206. Protection of human property							√			√		√		
		1207. Recreation/tourism							√			√		√		
		1208. Cultural/spiritual activities												√		
		1209. Aesthetic appreciation										√				
		1210. Information, science, education, and research										√		√		
		1299. Other in-situ use												√		
2. Non-Use	21. Existence	2101. Existence												√	√	√
	22. Bequest	2201. Bequest												√	√	√
	29. Other non-use	2901. Other non-use												√	√	√

101

Table 4-9. Example of a NESCS Table Relating NESCS-S and NESCS-D Categories

Direct Use/Non-Use Class	Direct Use/Non-Use Subclass	Direct Use/Non-Use Detail	1. Water	2. Flora	3. Fauna	4. Other Biotic Components	5. Atmospheric Components	6. Soil	7. Other Abiotic Components	8. Composite End-Products	9. Other End-Products
1. Direct Use	11. Extractive Use	1101. Raw material for transformation	✓	✓	✓	✓		✓	✓		
		1102. Fuel/energy	✓	✓			✓	✓			
		1103. Industrial processing	✓	✓	✓		✓	✓			
		1104. Distribution to other users	✓	✓	✓			✓			
		1105. Support of plant or animal cultivation	✓	✓	✓						
		1106. Support of human health and life or subsistence		✓	✓	✓			✓		
		1107. Recreation/tourism	✓	✓	✓	✓			✓		
		1108. Cultural/spiritual activities	✓	✓							
		1109. Information, science, education, and research	✓	✓							
		1199. Other extractive use	✓		✓	✓			✓		
	12. In-situ Use	1201. Energy	✓				✓	✓			
		1202. Transportation medium	✓				✓	✓			
		1203. Support of plant or animal cultivation	✓				✓	✓			
		1204. Waste disposal/assimilation	✓					✓			
		1205. Protection or support of human health and life								✓	
		1206. Protection of human property									
		1207. Recreation/tourism	✓	✓	✓	✓			✓	✓	
		1208. Cultural/spiritual activities	✓	✓	✓	✓			✓	✓	
		1209. Aesthetic appreciation	✓	✓	✓		✓		✓	✓	
		1210. Information, science, education, and research	✓	✓	✓	✓			✓	✓	
		1299. Other in-situ use	✓							✓	

(continued)

Table 4-9. Example of a NESCS Table Relating NESCS-S and NESCS-D Categories (continued)

Direct Use/Non-Use Class	Direct Use/Non-Use Subclass	Direct Use/Non-Use Detail	1. Water	2. Flora	3. Fauna	4. Other Biotic Components	5. Atmospheric Components	6. Soil	7. Other Abiotic Components	8. Composite End-Products	9. Other End-Products
2. Non-Use	21. Existence	2101. Existence	✓	✓	✓	✓	✓	✓	✓	✓	✓
	22. Bequest	2201. Bequest	✓	✓	✓	✓	✓	✓	✓	✓	✓
	29. Other non-use	2901. Other non-use	✓	✓	✓	✓	✓	✓	✓	✓	✓

103

Next, this cross-walk can help to identify all of the mutually exclusive **pathways** linking specific changes in ecosystems (ΔN) or ecological end-products (ΔE) with specific uses, and thereby with changes in human well-being (ΔU) (see Figure 4-5). Thus, this cross-walk defines and identifies an FFES.

For example, consider a policy that primarily increases salmon runs in a river basin. The salmon in the river and the ocean, and the species that consume them, are the main ecological end-products of interest, each of which can be assigned a separate column in the table. Going down each column, one can then define the relevant use categories that are affected by the increase. For example, uses of river salmon might include recreation, cultural activities, and non-use values, whereas ocean salmon would also include distribution to other uses (e.g., commercial harvesting).

Finally, once these pathways have been established, the table provides an accounting platform for values attached to changes in ecosystem services. That is, the table cells can be used to store value estimates for individual pathways, which can then be aggregated across rows and columns (or both). For example, the cell representing the combination of river salmon (column) and tribal cultural activities (row) would contain an estimate of the total value added by the additional river salmon to all of the households that engage in the affected cultural activities.

It is important to note here that values of FFES typically vary with location, scale and time and these must be accounted for when quantifying and valuing changes in FFES. However, these attributes are not necessary for *classifying* FFES and are thus not a part of the NESCS 15-digit system.[67] It would also be challenging to develop mutually exclusive and exhaustive categories for FFES if location, scale and time attributes were to be included in the classification system.

4.4   Summary of the NESCS Structure

To summarize, the primary purpose of NESCS is to support welfare estimation of policy induced changes in ecosystems by identifying ways in which people directly use or appreciate outcomes provided by nature. This identification of pathways linking ecosystems to human uses provides the basis for then quantifying and valuing ecosystem services. NESCS provides a consistent conceptual framework and a classification system for systematically linking ecological systems that produce ecosystem services with human systems that directly use these services (i.e., market production systems and households).

---

[67]   Other classification systems do not include categories that account for location, scale and time attributes either.

The classification system uses a 4-group structure and a coding system to define mutually exclusive and exhaustive categories for linking ecosystem outcomes to direct human uses (i.e., identifying FFES). Figure 4-9 illustrates how the four groups fit into the pathway between policy changes and human welfare. Implementing the NESCS framework involves identifying the point of hand-off from the ecosystem to human beings, and identifying ways in which end-products are used by human beings. The structure of NESCS-S supports the first step by defining a classification structure for the two groups—Environment, and End-Products provided by them. NESCS-D supports the second step by defining a classification structure for the two groups Direct Use/Non-Use and Users.

Figure 4-9. Pathway Linking Policy Changes to Human Well-Being

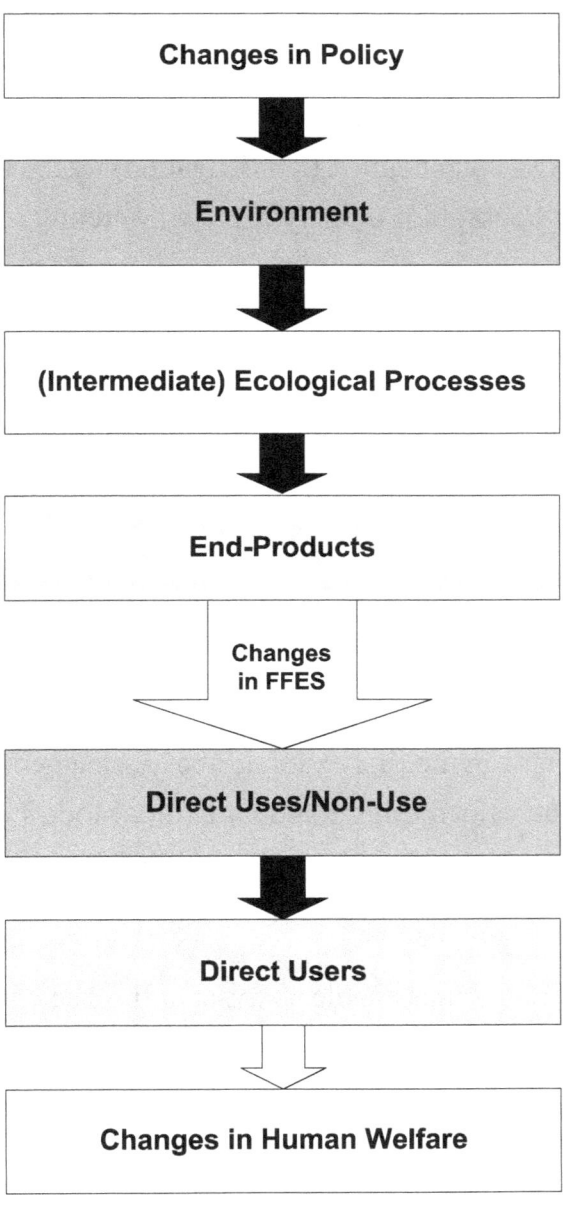

Flows of final ecosystem services (FFES) are represented by the connection from end-products to the human uses, that is, from the second group to the third group. Each environment–end-product–use–user combination thus identifies a potential FFES category and a unique pathway for linking policy changes with human welfare. Each unique FFES can be referenced by a NESCS code of up to 15 digits. Different combinations identify multiple mutually exclusive pathways, allowing the NESCS structure to be both flexible and comprehensive:

1. The same end-product can be used in multiple ways (e.g., water can be used to support human life [drinking water] and as an energy source [hydropower production]).

2. The same use can be linked to different sectors. For example, recreational uses can benefit households directly (recreational anglers), or benefit production processes in the transportation sector (tourism and sightseeing). Another example would be water being used to support plant cultivation (irrigation) by the agricultural sector for crop production, or by households directly for lawn watering.

This distinction between the use and user has been designed to provide flexibility to the analyst in the following ways:

1. Potentially, different values can accrue to different types of users from the same use. Thus, in the second example (irrigation) described above, the value to a commercial farmer may be different than for a household. The goal of NESCS is to identify pathways in a way that will support valuation; therefore we distinguish between uses and users to allow for this.

2. Potentially, different techniques may be necessary for valuing changes in uses to different users. Thus, in the first example above, changes in recreational uses to households may be valued using Random Utility Models, while changes in recreational uses to the travel industry may be valued using production function models.

The four groups in NESCS thus enable analysts to link the changes in policy to changes in ecosystems to changes in human welfare. The following section (Section 5) provides additional and more detailed illustrations of how NESCS can be applied for policy analysis.

SECTION 5
APPLICATION OF NESCS TO POLICY ANALYSES

5.1   Introduction

Using the NESCS conceptual framework and classification system described in Section 4, in this section we use two policy-related examples to illustrate how NESCS can be implemented to identify the pathways through which policy changes can ultimately result in human welfare changes. Specifically, we demonstrate how the NESCS four-group classification structure and coding system can be applied to identify and reference unique FFES pathways. This process involves the following main components (also summarized in Table 5-1):

- Based on region-specific scientific evidence and information, identify the environmental classes/subclasses and corresponding end-product classes/subclasses (defined in Tables 4-2 and 4-3) that are likely to be impacted.

- For the affected environmental classes/subclasses, apply Table 4-9 to identify the specific combinations of end-products and direct uses (defined in Tables 4-5 and 4-6) that are likely to be impacted.

- Apply other tables and tools provided by NESCS tables (such as Table 4-8) to identify relevant user categories (defined in Table 4-7) that directly use the end-products.

To demonstrate the wide range of contexts in which NESCS can be applied, we consider two very different hypothetical policy applications. The first application (described in Section 5.2) is a policy that reduces atmospheric deposition of acidifying and nutrient enriching compounds, such as nitrogen oxides (NO_x) and sulfur oxides (SO_x). It is a policy that directly alters the quality of multiple environmental classes. We focus on the atmosphere as an example. These changes are assumed to occur on a national or large regional scale. The second application (described in Section 5.3) focuses on wetlands restoration. In this case, the direct policy impact can be characterized as a change in the quantity (i.e., stock) of natural capital in an environmental class—wetlands. These changes are assumed to occur on a local or small regional scale.

We emphasize that these examples are included for illustrative purposes and to demonstrate some of the ways in which the framework can be used. Building off the conceptual structure in Figure 4-5, they are intended to show how a specific policy action can be linked to a large number of potential FFES pathways. Some of the identified pathways are included mainly

107

to demonstrate the framework rather than to suggest that there are large impacts along these pathways. In addition, these examples are not intended to be fully comprehensive. They should not be interpreted as identifying *all* of the potential pathways (including spillover effects and feedback effects) between a policy action and the resulting impacts on human well-being.

Table 5-1. How to Apply the NESCS Structure to Identify and Represent Unique FFES Pathways for Policy Analysis

How to…	NESCS Tools
…describe FFES pathways that may potentially be impacted by a policy change in a systematic and consistent manner?	Use NESCS conceptual framework (Figure 4-3) as guide
…identify unique FFES pathways?	
1. Identify the environmental classes/subclasses and corresponding end-product classes/subclasses that are likely to be impacted based on region-specific scientific evidence and information.	• Classification of Environment (Table 4-2) • Classification of End-Products (Table 4-3) • End-products in Each Environmental Class (Table 4-5)
2. Identify the specific combinations of end-products and direct uses/non-uses that are likely to be impacted	• NESCS Table Linking End-Products and Direct Uses/Non-Uses (Table 4-9) • Classification of Direct Use/Non-Use (Table 4-6)
3. Identify relevant user categories that directly use the end-products that are likely to be impacted	• NESCS Table Linking Direct Uses/Non-Uses with Users (Table 4-8) • Classification of Direct User (Table 4-7)
…reference and illustrate FFES pathways in a readily understandable manner?	
1. Diagrammatically	Fill in NESCS conceptual framework with categories identified (See Figures 5-1 through 5-5 as examples)
2. Numerically	Use NESCS 15-digit coding system (Tables 4-1, 4-2, 4-3, 4-5, 4-6, 4-7, 4-8, and 4-9)
…provide a structure that can be used to store values obtained from elsewhere?	
1. Use tables that link each of the four groups to organize, store, and present values (monetized or otherwise) that are obtained from other sources	• End-products in Each Environmental Class (Table 4-5) • NESCS Table Linking End-Products and Direct Uses/Non-Uses (Table 4-9) • NESCS Table Linking Direct Uses/Non-Uses with Users (Table 4-8)

5.2   Application 1: Policies to Reduce Acid and Nutrient Deposition

This section focuses on hypothetical air quality regulations to reduce atmospheric deposition of acidifying and nutrient enriching compounds such as NO_x and SO_x. This policy was selected for its potential to impact human beings through multiple pathways, as illustrated in Figure 5-1.[68] First, on the NESCS-S side, the policy can impact multiple environmental classes and subclasses (e.g., lakes, streams, forests) as well as multiple end-products (e.g., fish, sugar maple trees, red spruce trees). Second, on the NESCS-D side, it can impact multiple direct uses (e.g., raw materials, recreation, aesthetic appreciation) and multiple direct users (e.g., Forestry and logging sector [NAICS code 113] and households).

Third, there are multiple ecological mechanisms or processes through which the same type of FFES can be impacted. For example, NO_x deposition can contribute to both the acidification and nutrient enrichment of surface waters. Therefore, the contributions of fish stocks to recreational fishing may be impacted through both ecological processes. The NESCS structure is flexible enough to consider these two mechanisms separately. It also allows the analyst to organize and present the combined impact of the two mechanisms on each FFES, if that is preferred. However, in what follows, we focus only on potential impacts of the policy through the *acidification* mechanism (illustrated in Figures 5-2 and 5-3).

Table 5-2 identifies the primary components of NESCS-S (i.e., environmental classes and end-products) that may potentially be impacted by acidification. Aquatic systems that may change include rivers, streams, lakes, and ponds, while the main terrestrial system that may change is forests. Among the end-products provided by aquatic environmental systems (see Table 4-5), fauna, specifically fish and waterfowl, may change. Specific species of flora such as red spruce and sugar maple trees provided by forests may be affected. These end-products are directly used or appreciated by humans.

---

[68]   By focusing specifically on the effects of reducing acid deposition, this example by design does not address reductions in human health risks from breathing cleaner air. However, a comprehensive CBA of this policy would need to consider these potential co-benefits as well.

Figure 5-1. Potential Multiple Pathways Linking NO_xSO_x Policy Changes to Welfare Changes

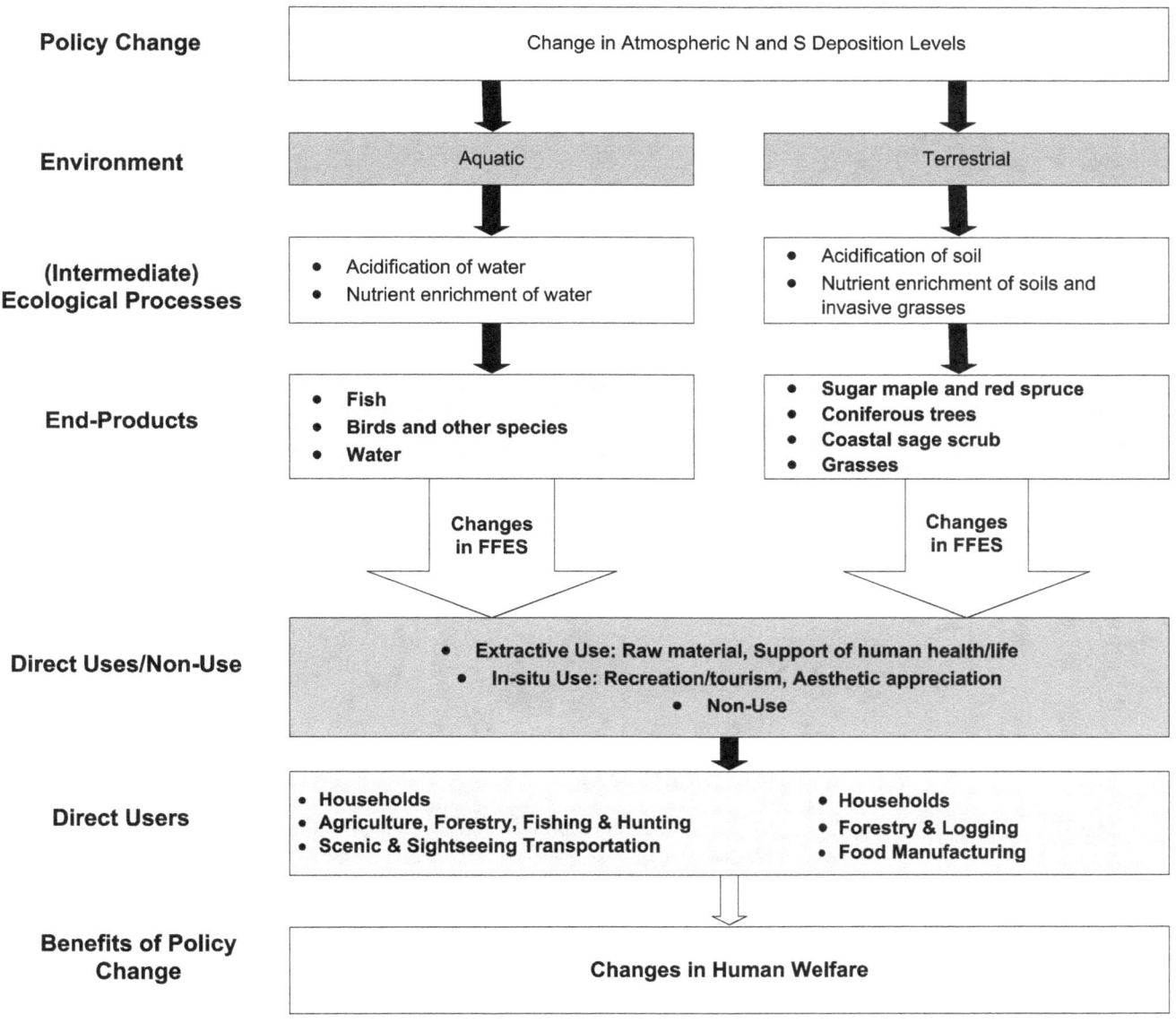

Figure 5-2. Applying the NESCS Framework: Identify Potential Pathways Impacted by Terrestrial Acidification

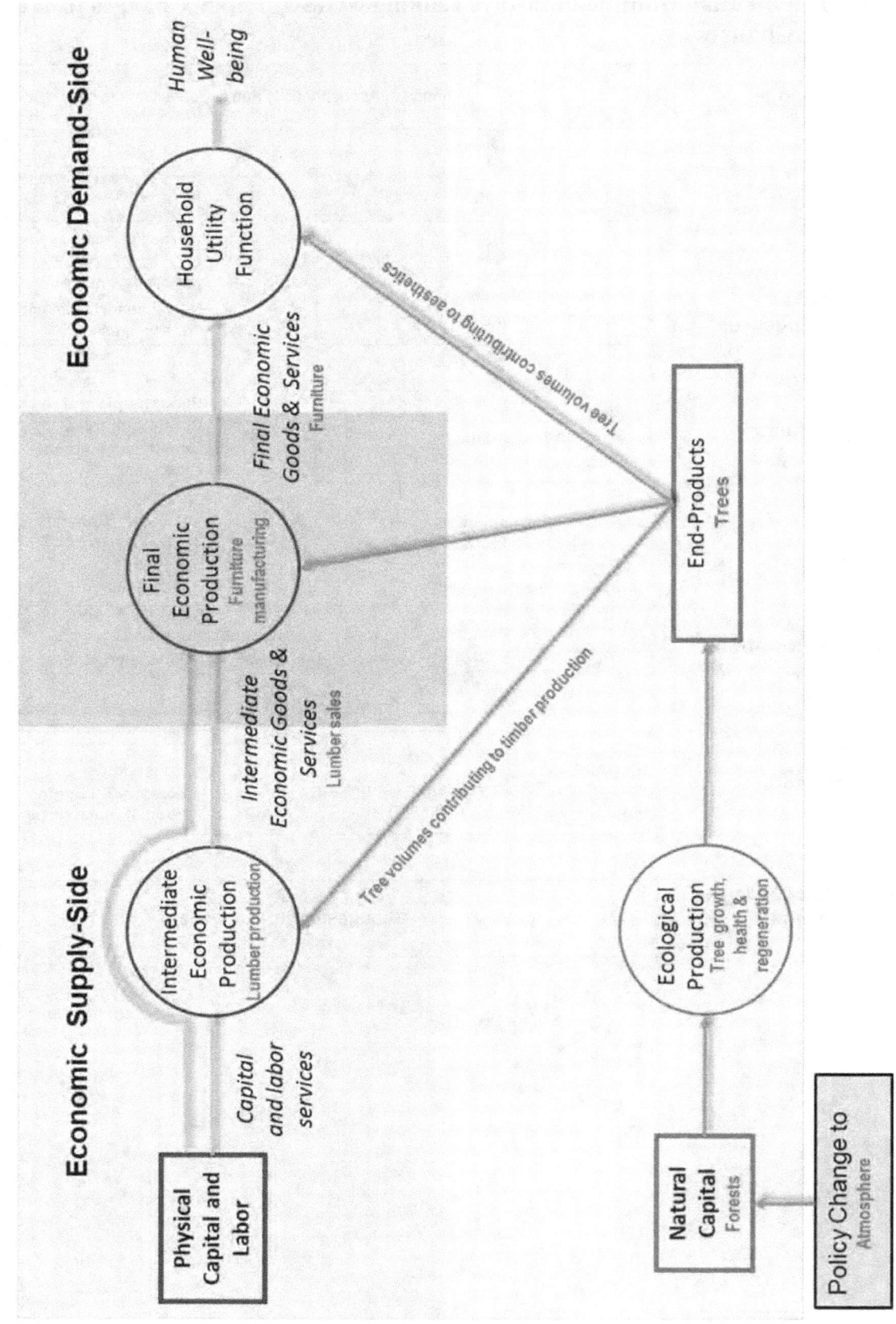

Figure 5-3. Applying the Framework: Identify Potential Pathways Impacted by Aquatic Acidification

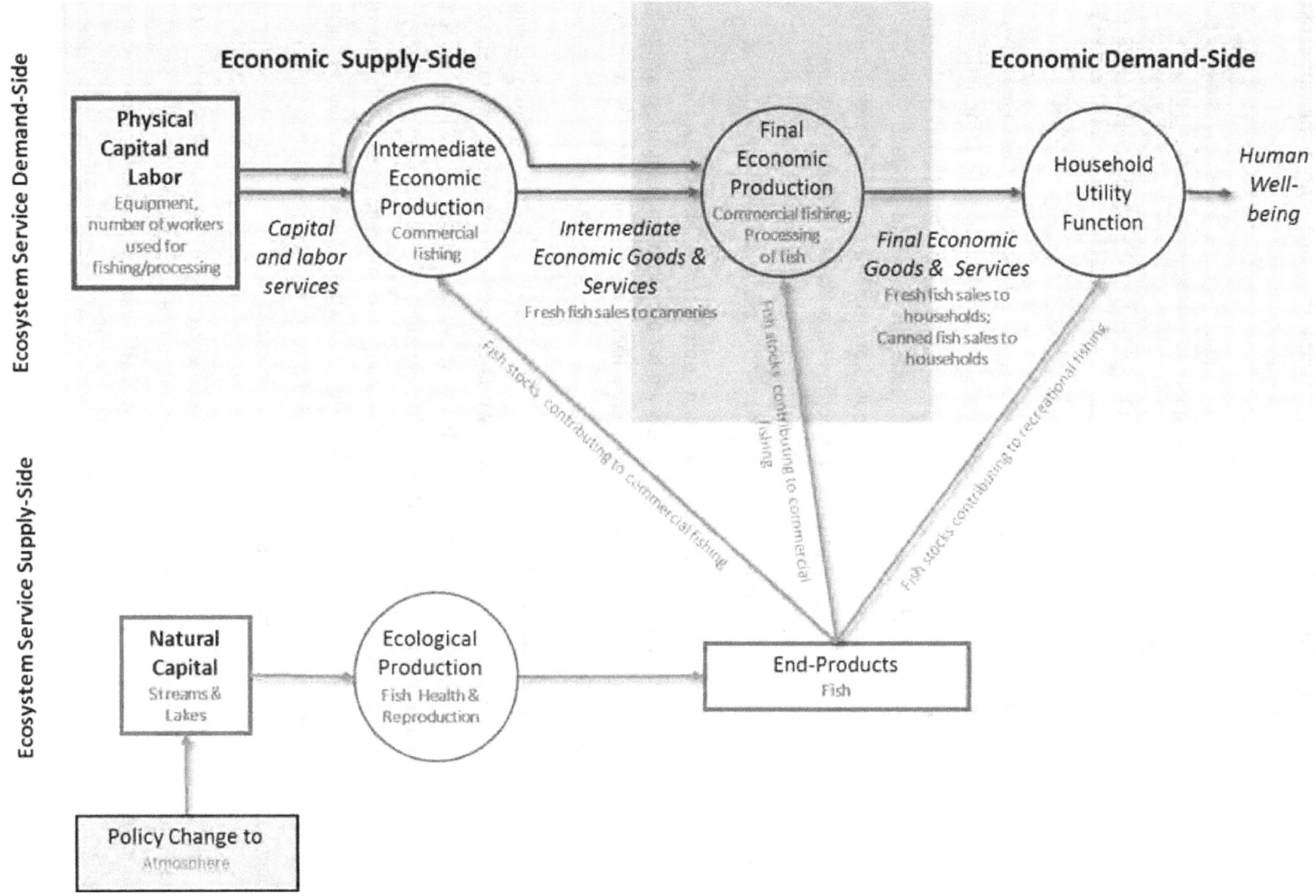

Table 5-2. **Environmental and End-Product Classes/Subclasses Likely to be Impacted by Acidification**

Environmental Class	Environmental Subclass	End-Product Class	End-Product Subclass	End-Product Subclass Examples
1. Aquatic	11. Rivers and Streams	3. Fauna	Specific classes/species of fauna	Fish
	13. Lakes and Ponds			Waterfowl
2. Terrestrial	21. Forests	2. Flora	Specific classes/species of flora	Sugar maple trees
				Red spruce trees

Table 5-3 illustrates how a table can be used as a tool for identifying linkages between end-products and uses. Such tables can be used in conjunction with scientific information and evidence to identify FFES that are likely to be impacted by a policy. For example, although water may be affected by acidification, we do not include this as an end-product. Scientific information indicates that direct uses of water are not likely to be impacted in a significant way due to reduced acidification resulting from this specific policy action. Water is of course critical for the ecological production of fish, waterfowl, and other wildlife, but in this role it is part of an intermediate process rather than the source of a final service of direct use to humans.

Tables 5-4 and 5-5 illustrate some of the main components of NESCS-D, that is, direct uses and users of the end-products that may be affected for terrestrial and aquatic systems respectively. Table 5-6 illustrates how direct use–direct user combinations that may be relevant for each impacted end-product can be identified.

For terrestrial systems (Table 5-4), both sugar maple and red spruce trees may be sources of non-use value to households. For our example, we assert that both extractive and in-situ uses of sugar maple trees may be affected, but that only extractive uses of red spruce trees are likely to be impacted by acidification. Households and different industrial sectors may use sugar maple for extractive or in-situ purposes. For example, sugar maple wood can be used for construction, and sap from sugar maple can be used for maple syrup production by the Food Manufacturing sector, which are both categorized as extractive uses; however, the unique autumn foliage associated with maple trees supports an in-situ aesthetic use.

Both sugar maple and red spruce can be used as raw materials, but not necessarily by the same industrial sectors. Both species can be used by Forestry and Logging, Wood Product Manufacturing and Furniture Manufacturing (3-digit NAICS sectors). However, only red spruce is widely used in Musical Instruments Manufacturing (a 6-digit NAICS sector). The NESCS codes that represent these different pathways are included in Tables 5-2 and 5-4. For example, the NESCS code representing the FFES pathway where trees in forests are used as raw material in the Forestry and Logging sector is 21.2.1101.113. On the other hand, trees in forests that are used as raw material in the Musical Instrument Manufacturing sector can be represented by 21.2.1101.1339992. The ability to use different NAICS aggregation levels for relevant users highlights the flexibility of the hierarchical system.

Although we have included inputs for maple syrup as an extractive use, the tree is still present after the sap is extracted. However, once the sap is extracted by someone, it is not available for extraction by others within a short period. Thus, within a certain time period, we include it as an example of an extractive use of the tree. After a certain length of time, the sap may be available for extraction again. This highlights the temporal nature of the FFES and is somewhat of a "gray" area for the classification system. This example also highlights the challenge in defining the end-product. Specifically, one must determine whether to represent the tree as the end-product or the sap, which is an attribute of the tree. As long as there is consistency about how an FFES is categorized, double counting and ambiguity issues can be avoided.

Both households and industrial sectors (e.g., Sightseeing and Scenic Transportation sectors and the Lodging and Accommodation sectors) directly derive benefits from recreation and tourism. This is an example of multiple users who derive value from the same use. Households who purchase tickets for scenic tours to view fall color foliage are "downstream" or indirect users. Including values derived by such households in addition to the industrial sectors mentioned would result in double counting. The code representing the pathway described above for Sightseeing and Scenic Transportation sectors is 21.2.1207.1487. However, households who visit the forests in the Northeast United States (for example) to take scenic tours (but do not purchase tickets) are direct users and this pathway can be represented by NESCS code 21.2.1207. 201. Another example of an in-situ use may be households who enjoy the aesthetics of fall colors on their daily commute (represented by NESCS code 21.2.1209.201). Thus, though the direct user category is the same (households), they derive values from different types of uses. It is important to distinguish between the two, as each has different implications for both data collection and determining appropriate methods for quantification and valuation. Therefore each has a distinct pathway in our example.

Table 5-5 provides examples of linkages between the end-products provided by aquatic systems and the appropriate NESCS-D categories. These examples also demonstrate that different users can derive value from the same use (e.g., recreation/tourism values derived from fish). It also shows that recreational uses can be both extractive and in-situ (e.g., catch-and-release fishing vs. catch-and-consume fishing, or waterfowl hunting vs. wildlife viewing).

Table 5-3. Tool to Identify Link Direct Uses/Non-Uses to End-Products

Direct Use/ Non-Use Class	Direct Use/Non-Use Subclass	Direct Use/Non-Use Detail	1. Water	2. Flora	3. Fauna	4. Other Biotic Components	5. Atmospheric Components	6. Soil	7. Other Abiotic Components	8. Composite End-Products	9. Other End-Products
1. Direct Use	11. Extractive Use	1101. Raw material for transformation	✓	✓	✓	✓		✓	✓		
		1102. Fuel/energy	✓	✓							
		1103. Industrial processing	✓	✓	✓		✓	✓			
		1104. Distribution to other users	✓	✓	✓		✓	✓			
		1105. Support of plant or animal cultivation	✓	✓	✓						
		1106. Support of human health and life or subsistence	✓	✓	✓						
		1107. Recreation/tourism	✓	✓	✓	✓			✓		
		1108. Cultural/spiritual activities	✓	✓	✓	✓			✓		
		1109. Information, science, education, and research	✓	✓	✓	✓			✓		
		1199. Other extractive use									
	12. In-situ Use	1201. Energy	✓				✓				
		1202. Transportation medium	✓				✓	✓			
		1203. Support of plant or animal cultivation	✓		✓		✓	✓		✓	
		1204. Waste disposal/assimilation	✓				✓	✓		✓	
		1205. Protection or support of human health and life	✓					✓		✓	
		1206. Protection of human property	✓							✓	
		1207. Recreation/tourism	✓	✓	✓	✓			✓	✓	
		1208. Cultural/spiritual activities	✓	✓	✓	✓			✓	✓	
		1209. Aesthetic appreciation	✓	✓	✓	✓			✓	✓	
		1210. Information, science, education, and research	✓	✓	✓	✓			✓	✓	
		1299. Other in-situ use	✓							✓	
2. Non-Use	21. Existence	2101. Existence	✓	✓	✓	✓	✓	✓	✓	✓	✓
	22. Bequest	2201. Bequest	✓	✓	✓	✓	✓	✓	✓	✓	✓
	29. Other non-use	2901. Other non-use	✓	✓	✓	✓	✓	✓	✓	✓	✓

Table 5-4. Direct Uses/Users Likely to be Impacted by Terrestrial Acidification

End-Product Subclass or Example	Direct Use/Non-Use Class	Direct Use/Non-Use Subclass	Direct Use/Non-Use Detail	Examples of Direct Uses/Non-Use	Direct User Class	Direct User Subclass	User Detail
Sugar maple trees	1. Direct Use	11. Extractive Use	1101. Raw material for transformation	Input for maple syrup, furniture, construction	1. Industry	111. Agriculture, Forestry, Fishing and Hunting	1113. Forestry and Logging
						123. Construction	123. Construction
						131–33. Manufacturing	1311. Food Manufacturing 1321. Wood Product Manufacturing 1337. Furniture and Related Product Manufacturing
		12. In-situ Use	1207. Recreation/ tourism	Fall color viewing	1. Industry	148–49. Transportation and Warehousing	1487. Scenic and Sightseeing Transportation
						172. Accommodation and Food Services	1721. Accommodation 1722. Food Services and Drinking Places
					2. Households	201. Households	
			1209. Aesthetic appreciation	Scenic views for commuters	2. Households	201. Households	
	2. Non-Use	21. Existence	2101. Existence	Existence use	2. Households	201. Households	
		22. Bequest	2201. Bequest	Bequest use	2. Households	201. Households	
Red spruce trees	1. Direct Use	11. Extractive Use	1101. Raw material for transformation	Input for musical instruments, furniture, construction	1. Industry	111. Agriculture, Forestry, Fishing and Hunting	1113. Forestry and Logging
						131–33. Manufacturing	1321. Wood Product Manufacturing 1337. Furniture and Related Product Manufacturing 1339992. Musical Instrument Manufacturing
	2. Non-Use	21. Existence	2101. Existence	Existence use	2. Households	201. Households	
		22. Bequest	2201. Bequest	Bequest use	2. Households	201. Households	

Table 5-5. Direct Uses/Users Likely to be Impacted by Aquatic Acidification

End-Product Subclass or Example	Direct Use/Non-Use Class	Direct Use/Non-Use Subclass	Direct Use/Non-Use Detail	Examples of Direct Uses/ Non-Use	Direct User Class	Direct User Subclass	User Detail
Fish	1. Direct use	11. Extractive Use	1104. Distribution to other users	Commercial fishing	1. Industry	111. Agriculture, Forestry, Fishing and Hunting	111. Agriculture
			1106. Support of human health or life or subsistence	Subsistence fishing	2. Households	201. Households	
			1107. Recreation/tourism	Recreational fishing	1. Industry	148–49. Transportation and Warehousing	1487. Scenic and Sightseeing Transportation
							171393. Marinas
						171. Arts, Entertainment, and Recreation	
						172. Accommodation and Food Services	1721. Accommodation 1722. Food Services and Drinking Places
					2. Households	201. Households	
		12. In-situ Use	1207. Recreation/tourism	Catch and release fishing	1. Industry	148–49. Transportation and Warehousing	1487. Scenic and Sightseeing Transportation
							171393. Marinas
						171. Arts, Entertainment, and Recreation	
						172. Accommodation and Food Services	1721. Accommodation 1722. Food Services and Drinking Places
					2. Households	201. Households	
	2. Non-Use	21. Existence	2101. Existence	Existence use	2. Households	201. Households	
		22. Bequest	2201. Bequest	Bequest use	2. Households	201. Households	

(continued)

Table 5-5. Direct Uses/Users Likely to be Impacted by Aquatic Acidification (continued)

End-Product Subclass or Example	Direct Use/Non-Use Class	Direct Use/Non-Use Subclass	Direct Use/Non-Use Detail	Examples of Direct Uses/Non-Use	Direct User Class	Direct User Subclass	User Detail
Waterfowl	1. Direct use	11. Extractive Use	1104. Distribution to other users	Commercial hunting	1. Industry	111. Agriculture, Forestry, Fishing and Hunting	111. Agriculture
			1106. Support of human health and life or subsistence	Subsistence hunting	2. Households	201. Households	
			1106. Support of human health and life or subsistence	Waterfowl hunting	1. Industry	148–49. Transportation and Warehousing	1487. Scenic and Sightseeing Transportation
						171. Arts, Entertainment, and Recreation	171393. Marinas
						172. Accommodation and Food Services	1721. Accommodation 1722. Food Services and Drinking Places
					2. Households	201. Households	
		12. In-situ Use	1107. Recreation/tourism	Wildlife viewing	1. Industry	148–49. Transportation and Warehousing	1487. Scenic and Sightseeing Transportation
						171. Arts, Entertainment, and Recreation	171393. Marinas
						172. Accommodation and Food Services	1721. Accommodation 1722. Food Services and Drinking Places
					2. Households	201. Households	
	2. Non-Use	21. Existence	2101. Existence	Existence use	2. Households	201. Households	
	2. Non-Use	22. Bequest	2201. Bequest	Bequest use	2. Households	201. Households	

Table 5-6. Tool to Identify Linkages between Direct Uses/Non-Uses and Direct Users

Direct Use/Non-Use Class	Direct Use/Non-Use Subclass	Direct Use/Non-Use Detail	111. Agriculture, Forestry, Fishing and Hunting	121. Mining	122. Utilities	123. Construction	131-33. Manufacturing	142. Wholesale Trade	144-45. Retail Trade	148-49. Transportation and Warehousing	151. Information	152. Finance and Insurance	Total market sector use	201. Households	301. Government	Total non-market sector use
1. Direct Use	11. Extractive Use	1101. Raw material for transformation	✓			✓	✓	✓						✓		
		1102. Fuel/energy			✓		✓							✓		
		1103. Industrial processing		✓	✓	✓	✓	✓	✓							
		1104. Distribution to other users	✓													
		1105. Support of plant or animal cultivation	✓													
		1106. Support of human health and life or subsistence														
		1107. Recreation/tourism	✓											✓		
		1108. Cultural/spiritual activities					✓		✓					✓		
		1109. Information, science, education, and research														
		1199. Other extractive use														

(continued)

120

Table 5-6. Tool to Identify Linkages between Direct Uses/Non-Uses and Direct Users (continued)

Direct Use/Non-Use Class	Direct Use/Non-Use Subclass	Direct Use/Non-Use Detail	111. Agriculture, Forestry, Fishing and Hunting	121. Mining	122. Utilities	123. Construction	131–33. Manufacturing	142. Wholesale Trade	144–45. Retail Trade	148–49. Transportation and Warehousing	151. Information	152. Finance and Insurance	Total market sector use	201. Households	301. Government	Total non-market sector use
1. Direct Use (cont.)	12. In-situ Use	1201. Energy			✓									✓		
		1202. Transportation medium							✓					✓		
		1203. Support of plant or animal cultivation	✓											✓		
		1204. Waste disposal/assimilation	✓	✓	✓	✓	✓	✓	✓	✓	✓	✓		✓		
		1205. Protection or support of human health and life							✓			✓		✓		
		1206. Protection of human property							✓					✓		
		1207. Recreation/tourism							✓			✓		✓		
		1208. Cultural/spiritual activities										✓				
		1209. Aesthetic appreciation														
		1210. Information, science, education, and research										✓		✓		
		1299. Other in-situ use										✓		✓		
2. Non-Use	21. Existence	2101. Existence												✓	✓	✓
	22. Bequest	2201. Bequest												✓	✓	✓
	29. Other non-use	2901. Other non-use												✓	✓	✓

The examples above demonstrate the design features of NESCS that play an important role in identifying distinct policy-relevant pathways that will support quantification and valuation. In particular, it demonstrates the flexibility in being able to use the appropriate environmental-class–end-product–direct-use–direct-user combinations, depending on the context and desired aggregation level.

5.3   Application 2: Wetland Restoration Policies

To further illustrate how NESCS can be used to identify discrete FFES pathways linking a policy action to human welfare, in this section we consider a hypothetical wetland restoration program. In particular, we use the example of a policy action involving the conversion to freshwater wetlands of multiple acres of agricultural land along a river network. This example was selected for two main reasons. First, it contrasts with the previous air pollution control example, in that it focuses on a policy action targeted toward a change in land use (with indirect implications for environmental quality), rather than a policy action targeting environmental quality. Second, wetlands are always a useful example for illustrating FFES, due to the multiple ecological functions that wetlands perform.

To organize the presentation and discussion of this example, we begin by distinguishing between five main ecological functions that wetlands serve:

- groundwater recharge;

- surface water storage;

- water purification/filtration;

- wildlife habitat provision; and

- open space provision.

Each of these functions can be broadly represented as an ecological production process that either (1) directly generates one of more ecological end-products, or (2) provides inputs to other processes that produce these end-products.

It is worth noting that these functions are often described in the literature as wetland ecosystem services (e.g., Ramsar, 2011). We use the term "function" to emphasize that, whereas these processes are all important contributors to human well-being, they do not necessarily represent final ecosystem services.

Table 5-7. Example of NESCS-S Categories Associated with Five Wetland Functions

Wetland Function	Environmental Class	Environmental Subclass	End-Product Class	End-Product Subclass	End-Product Examples
Groundwater Recharge	1. Aquatic	16. Groundwater	1. Water	12. Liquid water	
Water Storage	1. Aquatic	12. Wetlands	8. Composite End-Products	82. Regulation of extreme events	Flood Surge Reduction
Water Purification	1. Aquatic	11. Rivers and Streams	1. Water	12. Liquid water	
		14. Near Coastal Marine	3. Fauna	Specific classes/ species of fauna	Fish
Wildlife Habitat Provision	1. Aquatic	12. Wetlands	3. Fauna	Specific classes/ species of fauna	Waterfowl
					Wading birds
Open Space Provision	1. Aquatic	12. Wetlands	8. Composite End-Products	81. Scapes: • views • sounds and scents of land, sea, sky or a combination	Wetland Landscape

Table 5-7 presents the main NESCS-S pathways for each of these selected wetland functions. In all five cases, FFES are mainly affected through changes in aquatic environments. The affected environmental classes include groundwater for the recharge function, rivers, streams, and estuaries for the water purification function, and wetlands themselves for the other functions.

A variety of affected ecological end-products are also highlighted across wetland functions and environmental classes, including water, fish, birds, flood surge control, and wetland landscapes. For the case of water purification, we assume that in-stream water is the main end-product of interest from affected rivers and streams. However, for downstream estuarine waters, we assume that changes in fish stocks due to improved water quality are the main end-product.

For selected wetland functions and end-products, Figures 5-4 and 5-5 use the conceptual framework described in Section 4 to illustrate specific FFES pathways linking the policy action to human well-being (i.e., through both NESCS-S and NESCS-D). Figure 5-4 focuses on the linkages associated with groundwater recharge. In this diagram, the recharge capacity provided by the restored wetlands is represented as the key ecological production process, and the resulting groundwater supplies are the main end-product of this process. The diagram also highlights three main examples of direct uses of these groundwater supplies, all of which are extractive uses. The first use is as a source of irrigation water for cultivating crops (by agricultural producers), the second use is as a raw material input for beverage production, and the third use is as a source of life support for households using private wells. The diagram also shows that households benefit indirectly from the first two uses through the beverages they purchase in the market.

Figure 5-5 focuses on the open space function of wetlands. In this particular example, the ecological production process is the production/provision of the wetland landscape, which is also the end-product. This "composite" end-product (e.g., including water, grasses, wading birds, etc.) is shown to support one main type of in-situ use—that is, as an aesthetic amenity—for two different categories of direct users. The first direct user is a hotel located in the restored wetland landscape. In this case, the hotel "uses" the beauty of the natural setting to enhance the lodging services it sells to its guests. The guests represent the indirect user households because they can only gain access to this scenery through the market purchase of the hotel's services. The second direct user is a household whose home is also located in the restored landscape, and the residents directly benefit from the aesthetic setting. Figures 5-6 and 5-7 provide illustration of FFES pathways associated with water purification and water storage.

Table 5-8. Examples of FFES Pathway Categories Associated with the Groundwater Recharge Function

End-Product Subclass or Example	Direct Use/ Non-Use Class	Direct Use/Non-Use Subclass	Direct Use/ Non-Use Detail	Examples of Uses/Non-Uses	Direct User Class	Direct User Subclass	User Detail
12. Liquid water	1. Direct Use	11. Extractive Use	1101. Raw material for transformation	Beverage production	1. Industry	131–33. Manufacturing	13121. Beverage Manufacturing
			1105. Support of plant or animal cultivation	Irrigation for crop production	1. Industry	111. Agriculture, Forestry, Fishing and Hunting	111. Agriculture, Forestry, Fishing, and Hunting
			1103. Industrial processing	Cooling water	1. Industry	122. Utilities	12211. Electric Power Generation, Transmission and Distribution
					1. Industry	131–33. Manufacturing	13311. Iron and Steel Mills and Ferroalloy Manufacturing
			1104. Distribution to other users	Distribution to commercial and household users	1. Industry	122. Utilities	122131. Water Supply and Irrigation Systems
			1106. Support of human health and life or subsistence	Tap water from private wells	2. Households	201. Households	
	2. Non-Use	22. Bequest	2201. Bequest	Bequest value for future generations	2. Households	201. Households	

Figure 5-4. Illustration of FFES Pathways Associated with the Groundwater Recharge Function

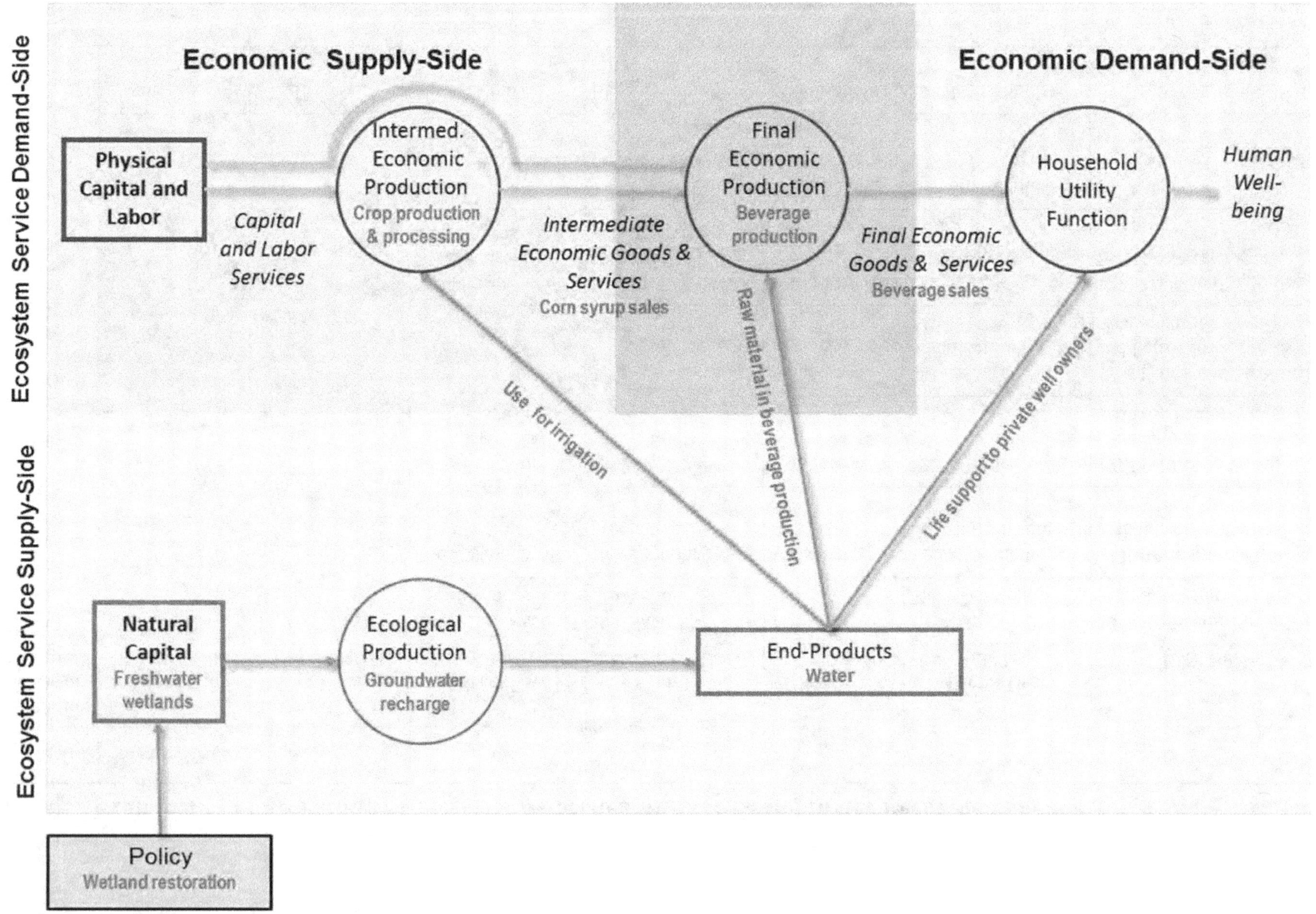

Figure 5-5. Illustration of FFES Pathways Associated with the Open Space Function

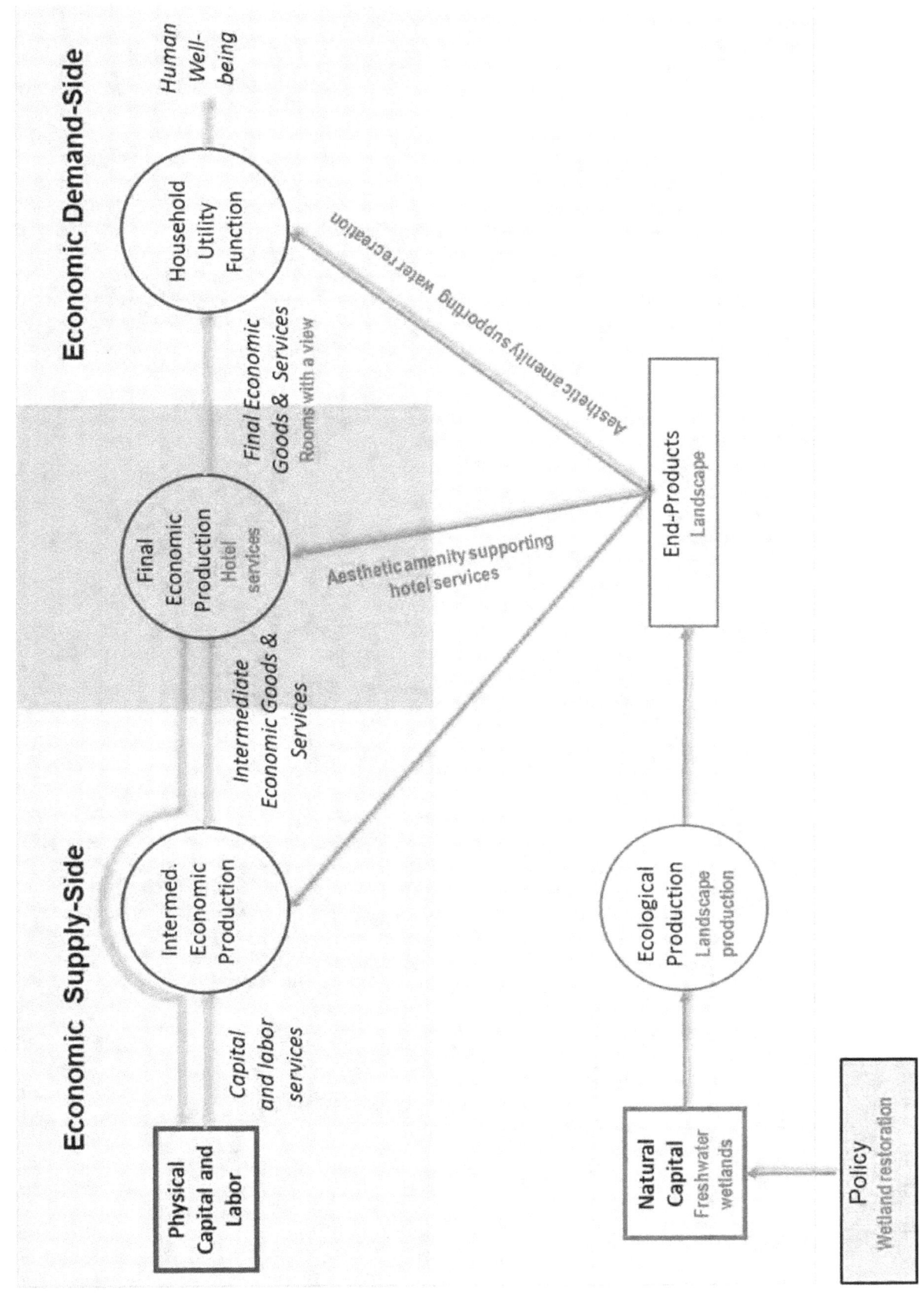

Figure 5-6. Illustration of FFES Pathways Associated with the Water Purification Function

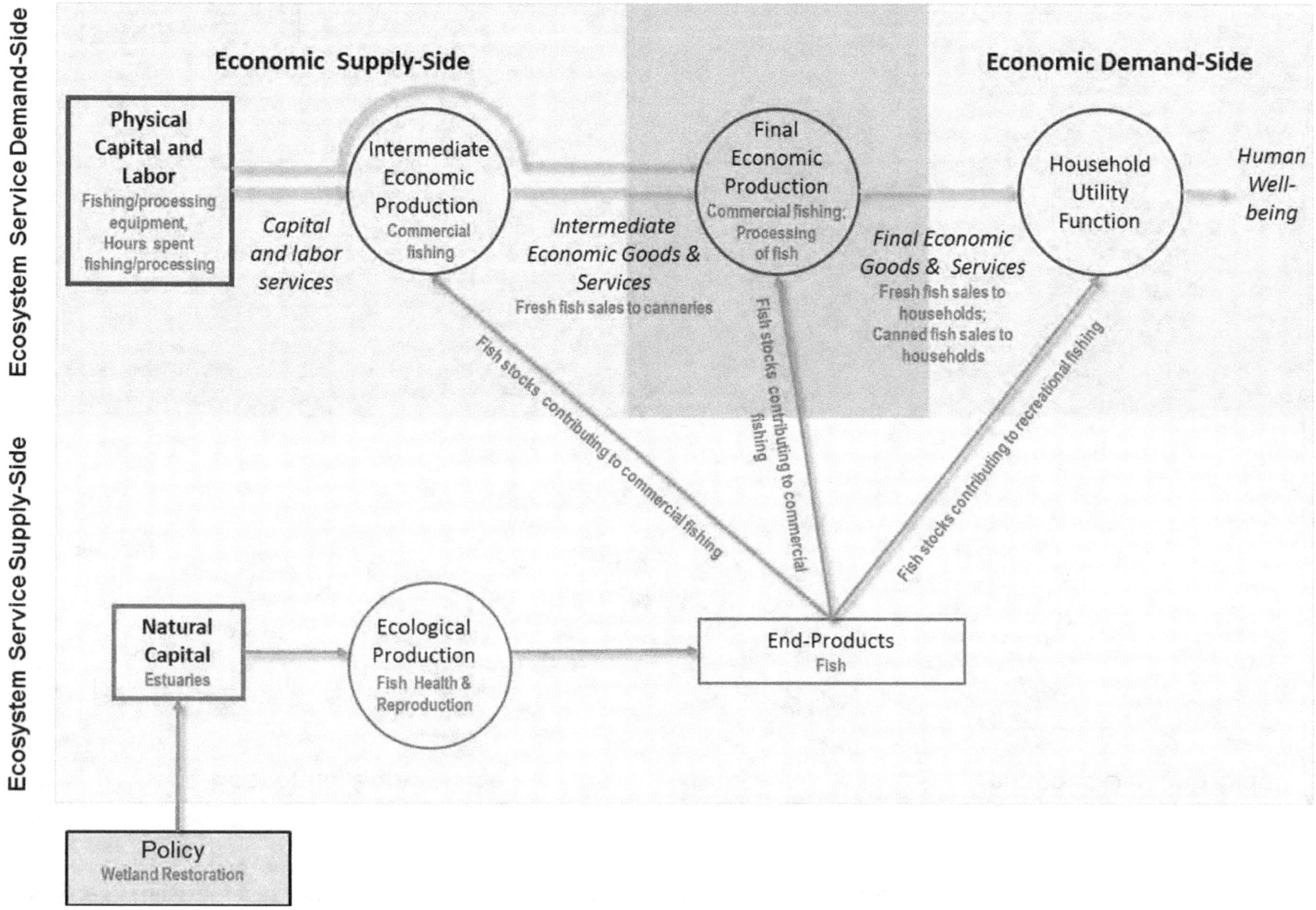

Figure 5-7. Illustration of FFES Pathways Associated with the Water Storage Function

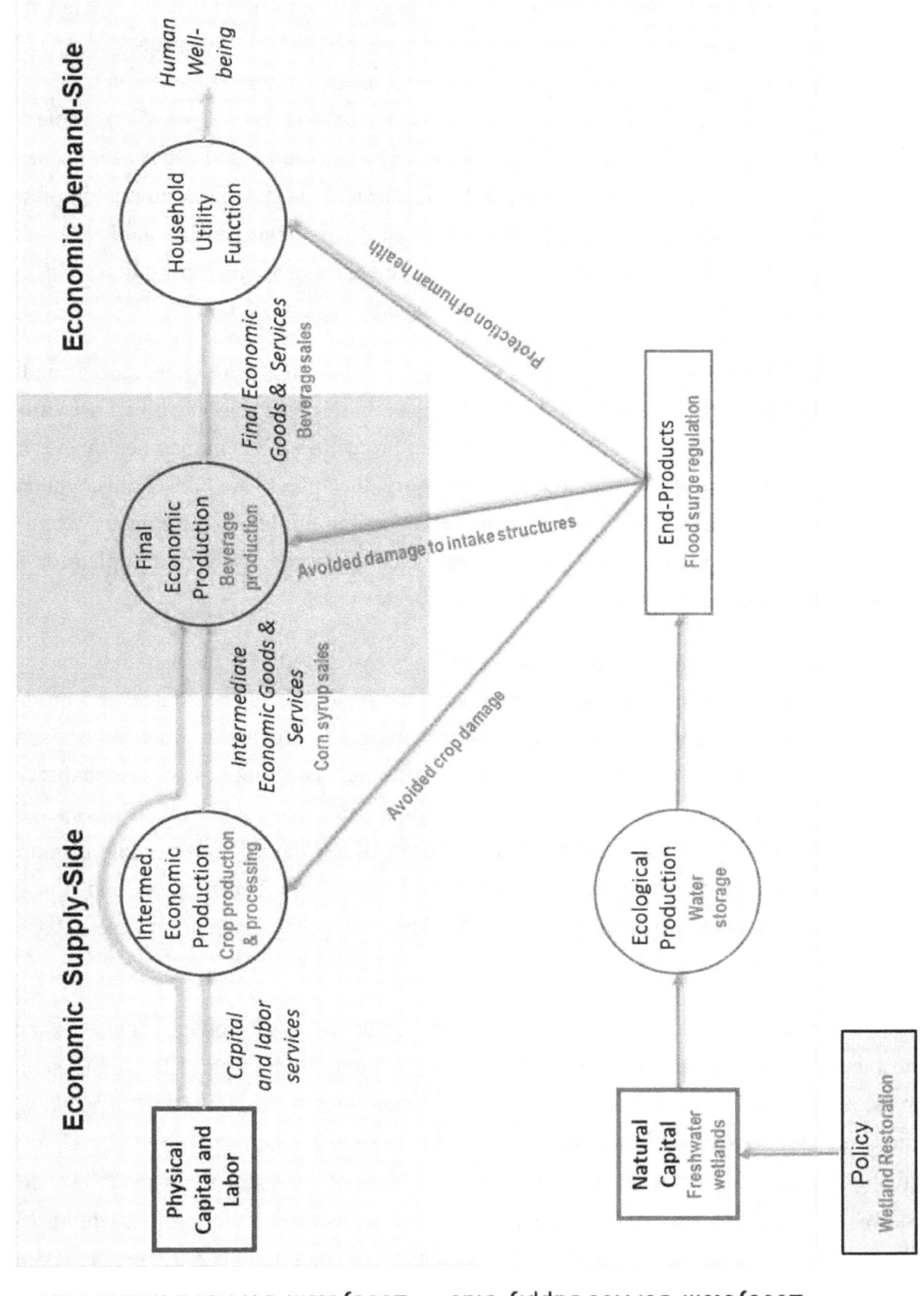

For each of the five wetland functions, Tables 5-8 to 5-12 provide a more detailed breakdown of hypothetical FFES pathways linking the NESCS-S end-products with specific direct use and user categories. For the groundwater recharge example, in addition to the irrigation, raw material, and life support uses described above (and in Figure 5-4), Table 5-8 identifies two additional direct uses of groundwater. The first additional use is as industrial processing (i.e., cooling) water for two separate industries. This specific example highlights how a single direct *use* category can apply to multiple direct *user* categories. This table also accounts for the possibility that households may derive non-use values (services) from the knowledge that groundwater resources are being restored and protected (bequest value).

Table 5-9 provides examples of FFES pathways associated with water storage and the resulting reduction in the size of periodic flood surges. In this example, the direct uses are all classified as in situ and are broadly divided between protection of (1) human health and life and (2) human property. For the health protection category, the "users" are households, whereas the property protection category includes both household and multiple market/industry sector users. This example also highlights how a single direct use category can apply to multiple user categories, including both household and industry sector users.

Table 5-10 identifies multiple FFES pathways associated with the wetlands water purification function. In nearby rivers and streams, the main affected ecological end-product is assumed to be in-stream water. This water has both extractive and in-situ uses and is also a source of non-use values. In this example, public drinking water supply systems are direct users of surface water resources, then they distribute the water to households and businesses. These customers are therefore considered to be indirect users of the water. To avoid double counting, they are not included as a separate user category in the table. In contrast, households that live along the affected rivers are included as direct users because they make in-situ use of the aesthetic amenities provided by surface water.

In this example, the water purification function provided by wetlands is also assumed to contribute to cleaner water and larger fish stocks in a downstream estuary. These fish stocks provide a variety of services through both extractive and in-situ uses by households and businesses. In this example, households that engage in recreational fishing by using chartered fishing boat services are not considered to be direct users of the resource. Instead the charter businesses are the direct users, who then sell recreational fishing services to households (who are indirect users). In contrast, households who use their own boats and gear are treated as direct users.

Table 5-9. Examples of FFES Pathway Categories Associated with the Water Storage Function

End-Product Subclass or Example	Direct Use/ Non-Use Class	Direct Use/Non-Use Subclass	Direct Use/ Non-Use Detail	Examples of Uses/Non-Uses	Direct User Class	Direct User Subclass	User Detail
Flood Surge Reduction	1. Direct Use	12. In-situ Use	1205. Protection or support of human health and life	Avoided drownings	2. Households	201. Households	
			1206. Protection of human property	Avoided crop damage	1. Industry	111. Agriculture, Forestry, Fishing and Hunting	111. Agriculture, Forestry, Fishing and Hunting
				Avoided damage to water intake structures	1. Industry	131–33. Manufacturing	13121. Beverage Manufacturing
				Avoided damage to vehicles	1. Industry	148–49. Transportation and Warehousing	1484. Truck Transportation
					2. Households	201. Households	
				Avoided residential damage	2. Households	201. Households	

131

Table 5-10. Examples of FFES Pathway Categories Associated with the Water Purification Function

End-Product Subclass or Example	Direct Use/Non-Use Class	Direct Use/Non-Use Subclass	Direct Use/Non-Use Detail	Examples of Uses/Non-Uses	Direct User Class	Direct User Subclass	User Detail
12. Liquid water	1. Direct Use	11. Extractive Use	1101. Raw material for transformation	Beverage production	1. Industry	131–33. Manufacturing	13121. Beverage Manufacturing
			1104. Distribution to other users	Distribution to commercial and household users	1. Industry	122. Utilities	122131. Water Supply and Irrigation Systems
		12. In-situ Use	1209. Aesthetic appreciation	Scenic amenity for waterside homes	2. Households	201. Households	
	2. Non-Use	22. Bequest	2201. Bequest	Bequest value for future generations	2. Households	201. Households	
Fish	1. Direct Use	11. Extractive Use	1104. Distribution to other users	Harvesting for sale by commercial fishers	1. Industry	111. Agriculture, Forestry, Fishing and Hunting	11141. Fishing
			1106. Support of human health and life or subsistence	Subsistence fishing	2. Households	201. Households	
			1107. Recreation/ tourism	Chartered recreational fishing	1. Industry	148–49. Transportation and Warehousing	1487210. Scenic and Sightseeing Transportation, Water
				Private recreational fishing	2. Households	201. Households	
		12. In-situ Use	1207. Recreation/ tourism	Catch-and-release private fishing	2. Households	201. Households	
	2. Non-Use	21. Existence	2101. Existence	Existence value	2. Households	201. Households	

132

Taken together, the water recharge, water storage, and water purification examples also highlight how a single direct user category—in this case beverage manufacturers—can be associated with multiple direct uses and FFES pathways. Due to these three wetland functions, they benefit from larger groundwater stocks, cleaner surface water sources, and avoided damage to their water intake structures.

Table 5-11 identifies FFES pathways associated with the provision of wetland wildlife habitat, in particular for waterfowl and wading birds. Unlike the estuarine fish end-product described above, these fauna end-products are assumed to exist within the wetland ecosystem. However, they support similar types of direct extractive and in-situ uses, including recreational and subsistence activities.

Expanding on the pathways shown in Figure 5-5, Table 5-12 identifies specific categories of direct uses and direct users of the wetland landscapes provided by the open space function. All of these uses are inherently in situ. In addition to the aesthetic appreciation use for nearby households and businesses, this example imagines recreational and ceremonial uses of the wetland landscape by certain households.

Table 5-11. Examples of FFES Pathway Categories Associated with the Wildlife Habitat Provision Function

End-Product Subclass or Example	Direct Use/Non-Use Class	Direct Use/Non-Use Subclass	Direct Use/Non-Use Detail	Examples of Uses/Non-Uses	Direct User Class	Direct User Subclass	User Detail
Waterfowl	1. Direct Use	11. Extractive Use	1106. Support of human health and life or subsistence	Subsistence hunting	2. Households	201. Households	
			1107. Recreation/ tourism	Waterfowl hunting preserves	1. Industry	111. Agriculture, Forestry, Fishing and Hunting	11142. Hunting and Trapping
				Private recreational hunting	2. Households	201. Households	
		12. In-situ Use	1207. Recreation/ tourism	Bird watching	2. Households	201. Households	
	2. Non-Use	21. Existence	2101. Existence	Existence value	2. Households	201. Households	
Wading birds	1. Direct Use	12. In-situ Use	1207. Recreation/ tourism	Birdwatching	2. Households	201. Households	
	2. Non-Use	21. Existence	2101. Existence	Existence value	2. Households	201. Households	

134

Table 5-12. Examples of FFES Pathway Categories Associated with the Open Space Function

End-Product Subclass or Example	Direct Use/ Non-Use Class	Direct Use/Non-Use Subclass	Direct Use/Non-Use Detail	Examples of Uses/Non-Uses	Direct User Class	Direct User Subclass	User Detail
Wetland landscape	1. Direct Use	12. In-situ Use	1208. Cultural/ spiritual activities	Ceremonies in wetland setting	2. Households	201. Households	
			1207. Recreation/ tourism	Canoeing/kayaking	2. Households	201. Households	
				Hiking	2. Households	201. Households	
			1209. Aesthetic appreciation	Scenic amenity for waterside homes	2. Households	201. Households	
				Scenic amenity for waterside businesses	1. Industry	172. Accommodation and Food Services	172111. Hotels (except Casino Hotels) and Motels
	2. Non-Use	22. Bequest	2201. Bequest	Bequest value for future generations	2. Households	201. Households	

SECTION 6
CONCLUSIONS

6.1   Summary of Report

Analyzing the human welfare impacts (benefits) of policy-induced changes in ecosystems typically entails identifying, quantifying, and valuing changes in ecosystems and their contributions to human welfare (EPA, 2009). However, ecosystems provide flows of services to humans through numerous and often complex pathways. The goal of NESCS is to provide a framework that helps identify these distinct pathways between natural and human systems. By helping to identify and classify these ecosystem service flows, it is also intended to support the quantification and valuation of ecosystem services. In this report we describe a conceptual framework for defining and identifying ecosystem services (in particular, flows of final ecosystem service), and we provide a classification structure and coding system based on this framework.

Section 2 reviews the literature on classification of ecosystem services. This section shows that although there is a common understanding that ecosystems support human welfare, there is disagreement on where ecosystem services occur along the continuum between ecosystems and human welfare. Boyd and Banzhaf (2007) addressed this issue by defining *final* ecosystem services as the end-products of nature, directly used or appreciated by humans. Most importantly, final ecosystem services occur at the point of hand-off between natural systems (ecosystems) and human systems (producers and households).

To define an approach for classifying ecosystem services, we adapt and apply some of the basic concepts and structures of the classification and accounting systems commonly used for economic goods and services (NAICS/NAPCS and NIPA). Section 3 describes these economic accounts and classification systems and their implications for the design of NESCS. In particular, it describes how the supply and demand side concepts from these systems are adapted for classifying ecosystem services.

The approach and methods for developing the NESCS system are described in Section 4. In addition to applying concepts from NAICS/NAPCS and NIPA, it adapts the logic and principles underlying the concept of final ecosystem services. Specifically, NESCS describes flows of final ecosystem services (FFES) as contributions that the end-products of nature provide directly to human production processes or directly to human well-being. Section 4 also introduces a conceptual framework that provides a way to organize and visualize the links between ecological systems and human systems. It also provides the foundation for the NESCS structure, which comprises four groups. NESCS-S includes two groups (environment and end-products), and NESCS-D includes the other two groups (direct use/non-use and direct users). Each potential and mutually exclusive FFES pathway is identified by a unique *combination* of the elements of these four groups (environment, end-product, direct use/non-use, and direct user) NESCS also provides an up to 15-digit coding system for numeric representation and referencing of FFES pathways.

In Section 5, we illustrate how NESCS serves its purpose with policy examples. Specifically, we demonstrate how the components of NESCS can be applied to identify potential pathways through which policy changes can ultimately result in changes in human welfare. We select as examples policies to change acid deposition and wetland restoration policies.

6.2   Key Features of NESCS

In summary, NESCS offers several key features for classifying ecosystem services.

First, it provides an explicit conceptual framework for defining FFES. This framework clearly distinguishes FFES (a) from the ecological production functions/processes that produce them; and (b) from the goods and services produced by human beings (particularly those requiring natural inputs, such as crops that require water and soil fertility). NESCS also defines *flows* of services to be consistent with standard economic concepts. To be consistent with the definition of services, NESCS explicitly separates out the supply-side (provider) and the demand-side (consumer) of FFES.

Second, it is designed to avoid double counting of ecosystem services. It does this by: (a) distinguishing between intermediate ecological production functions/processes and final ecosystem services; (b) striving to define mutually exclusive use categories; and (c) distinguishing between direct (e.g., fruit growers) and indirect users (e.g., households that consume fruit from growers). As described in previous sections, there will inevitably be "gray" areas where overlaps may exist; however, NESCS is intended to minimize those overlaps.

Third, NESCS provides a modular structure[69] intended to be as comprehensive and flexible as possible in capturing potential pathways from ecosystems to human beings. The hope is that this flexible structure will limit the need for extensive modifications to the classification system in the future. As new and unanticipated FFES become relevant in the future, it is intended

---

[69]   We allow for the fact that the same end-product can be used in multiple ways. For example, water can be used to support human life (as drinking water) and as an energy source (hydropower production). The same use can be linked to different sectors. For example, recreational uses can benefit households directly (recreational anglers), or benefit production processes in the transportation sector (tourism and sightseeing).

that they can be accomodated in NESCS by combining elements from the existing four groups. The modular structure in NESCS also provides flexibility to the analyst in conducting valuation in the following ways. Potentially, different values can accrue to different types of users from the same use. Thus, the value to a commercial farmer of changes in water may be different than for a household. Also, potentially, different techniques may be necessary for valuing changes in uses to different users. Thus, changes in recreational uses to households may be valued using Random Utility Models, while changes in recreational uses to the travel industry may be valued using production function models.

Fourth, NESCS leverages several existing methods and classification systems. In particular, it adapts and applies: (1) the FEGS-CS Environmental Classes, which are based on Anderson Land Use and Land Cover classes; (2) NAICS to define user categories; and (3) make and use table concepts from national economic accounts. In addition, to trace impacts from direct users in the market production sector (NESCS-D) to "downstream" consumers, it implicitly relies on NIPA input-output relationships.

Fifth, the NESCS framework captures causal links from ecosystems to human health and safety in a variety of ways. Although a distinction is often made between human health and other environmental benefits in discussions, analyses, and design of U.S. environmental policies,[70] the purpose of NESCS is to define, as comprehensively as possible, the pathways linking ecosystems to human well-being, including health-related pathways. First, it defines direct use categories for ecological end-products that have explicit connections to the health and safety of the direct users—for example, support and protection of human health and life. These categories include uses of air, water, nutrients, and natural hazard protections that are essential for human life. Second, NESCS defines direct use (and non-use) categories that may have indirect health effects on direct users. For example, individuals' physical and emotional health outcomes may be affected through direct recreational and other cultural uses of natural resources. Third, NESCS defines direct use (and non-use) categories that may have health implications for "downstream" users (i.e., buyers) of the economic goods and services produced through direct use of specific ecosystem inputs. For instance, purchases of food and medicine produced with natural inputs have health implications for consumers who are, in this case, indirect users of ecological end-products.

---

[70] For example, EPA's mission is stated as protecting "human health and the environment." The Clean Air Act (CAA) directs EPA to define separate national ambient air quality standards: (1) for protecting public health, and (2) for protecting against adverse effects on public welfare, including the deterioration of the quality of ecosystems. EPA's guidelines for conducting economic analyses of environmental policies (EPA, 2009) distinguish between two main categories of benefits—human health and ecological benefits.

To summarize NESCS, it is also important to emphasize its boundaries and areas that it does *not* cover. First, and most importantly, NESCS is *not an ecosystem service valuation system*. The goal of NESCS is to provide a framework for *identifying* categories of FFES that may be affected by policy-induced changes to ecosystems. In this way, it can help to organize and lay the groundwork for quantifying and valuing these changes, but it does not provide a system for estimating or calculating these changes.

Second, unlike the NIPA and SEEA systems described in Section 3, NESCS is not an accounting system. However, as discussed in more detail in Section 6.4, it may provide a useful framework for helping to organize national environmental accounts, including green GDP accounting.

Third, by design, the NESCS conceptual framework (as represented, for example, by Figure 4-3) does not define and categorize feedback effects (flows) from human systems to natural systems. For example, it does not include arrows representing how pollution or resource depletion from human activities affects ecosystems. Figure B-2 in Appendix B illustrates such effects. This omission is not intended to diminish the importance of these flows since these types of feedback effects must be considered when conducting comprehensive economic or resource accounting analyses. Feedbacks will generate more flows through the NESCS system and therefore more FFES pathways will need to be considered. However, considering these feedbacks does not imply that new FFES pathways will need to be *defined* and *classified*.

6.3   Comparison of NESCS with NAICS/NAPCS and FEGS-CS

Although NESCS was designed using concepts from the NIPA framework and the NAICS/NAPCS classification systems, in this section, we highlight the major distinctions between NESCS and NAICS/NAPCS. We also compare and contrast the NESCS system with the FEGS-CS.

One way in which the NESCS differs from the NAICS and NAPCS is that, whereas NAICS and NAPCS can be used to classify both *intermediate* and *final economic goods* and services, the NESCS structure focuses specifically on flows of *final ecosystem services*. For example, all output from crop production, whether used as an intermediate input in food manufacturing or as a final good sold to households, is classified under NAICS code 111. In contrast, although recognizing their importance for human well-being, NESCS does not include classification systems for ecological production processes or for the inputs to these processes, which can be thought of as intermediate ecosystem services. For example, nutrient cycling is essential for human life, but it is not separately classified in NESCS, because the value of this "service" is embedded in the soil and/or water end-products provided by nature. Just as the input-output relationships in the economy are left to other frameworks (i.e., NIPA), these intermediate ecosystem services and processes are assumed to be captured in other ecological models and frameworks.

Second, NESCS classifies the flow of final services derived from ecosystems that serve as *inputs* to economic production functions, while NAICS/NAPCS classifies the *output* from these economic production processes.

A third important distinction to note is that while NAICS and NAPCS provide alternative ways for classifying economic goods and services,[71] NESCS-S and NESCS-D together constitute the classification system for FFES. They are complementary systems that need to be used in conjunction with each other in order to identify and classify FFES.

A final key distinction is that NAICS and NAPCS primarily include goods and services that are produced and sold in markets, whereas NESCS primarily addresses ecosystem services which are not produced or sold in markets. One implication of a market system is that it provides incentives for producers to specialize in certain production activities. This tendency toward specialization does not mean that all producers only produce one type of good or service, but it does make it much easier to define a firm's *primary* production activity, which can then be used to categorize establishments according to NAICS categories. In contrast, ecosystems have less of a tendency to specialize in specific ecosystem services, because they are not motivated by market incentives. This lack of specialization implies that the NAICS categorization approach cannot be directly replicated for ecosystem service production in NESCS-S. A second implication of a market system is that it involves explicit transactions between producers and buyers. These terms of transactions (including the agreed-upon price) provide important information about the commodity (good or service) being exchanged, which can be used to define and categorize commodities using NAPCS. Because the provision of ecosystem services does not involve explicit market transactions, defining the relevant "commodity" in NESCS is inherently more difficult. Instead, it must be inferred based on how and by whom the ecosystem is being used.

NESCS and FEGS-CS also have important features in common. The ultimate purpose behind both systems is to provide a classification system for final ecosystem services that helps to inform environmental and natural resource policy and management decisions. The two approaches are based on the same fundamental conceptual framework for linking ecosystems to human welfare. However, there are differences between the two approaches in their specific

---

[71] Of course, NAICS and NAPCS together help support the accounting system (NIPA); however, they can be viewed as independent classification systems for economic goods and services.

objectives, methods, and organizational structures. We highlight some of the key similarities and differences between the two approaches in Table 6-1.

Table 6-1. Comparison of NESCS and FEGS-CS

	NESCS	FEGS-CS
Broad Goal	The goal of both approaches is to provide a classification system that is based on a "rigid framework in which ecosystem services can be identified on the landscape and explicitly associated with people." (p2, Landers and Nahlik, 2013)	
Key Features	Both approaches contain the following key features: • provide a standardized and consistent framework to promote communication and collaboration between natural and social sciences • connect ecosystems (components and processes) to human well-being • seek to avoid disconnects between the ecosystem components measured by ecologists and those valued by the public • seek to avoid double counting ecosystem services by focusing on the "final" end-products of nature that people directly care about • distinguish between the services that ecosystems provide to humans and the human uses and benefits that are supported by these services • identify ecosystem service categories by combining separate classification systems for (1) the environmental components that provide the services, and (2) the ways in which humans use and benefit from these components	
Specific Objectives and Focus	Develop a classification system that: (1) comprehensively and uniquely (without duplication) identifies distinct categories of final ecosystem services; (2) supports analysis of how policy-related changes in ecosystems affect human well-being.	Develop a classification system that will "determine those specific **ecosystem attribute(s)** associated with the specific FEGS that the beneficiary values" such that "these can directly lead to identifying appropriate **metrics and indicators** for FEGS" (p6, Landers and Nahlik, 2013).
Design and Implementation Approach	Develop the system by applying, adapting, and combining the principles underlying: (1) existing economic classification and accounting systems for market goods and services; (2) the concept of "final" ecosystem services described in Boyd and Banzhaf (2007)	FEGS are explicitly defined by the landscape in which they occur (Environmental Class) and the interests of the people that interact with the FEGS (Beneficiary Categories).
Definition of Final Ecosystem Services	<u>Flows of final ecosystem services (FFES)</u> are the contributions that the end-products of nature provide directly to human production processes or directly to human well-being. They are thus represented by service *flows* between ecological end-products and direct human uses.[72]	<u>Final ecosystem goods and services (FEGS[73])</u> are "components of nature, directly enjoyed, consumed or used to yield human well-being" (Boyd and Banzhaf, 2007). As a result, FEGS are more like a *stock* concept.[74]

(continued)

---

[72] To deliberately separate "stock" and "flow" concepts, the NESCS framework does not include the term "ecosystem goods." Instead, it uses the term "ecosystem end-products" to represent the stocks provided by nature and "ecosystem services" to represent the flows provided by nature.

[73] The concept of "final" ecosystem services was developed by Boyd and Banzhaf (2007) and EPA adopted the term FEGS later to represent this concept.

[74] Landers and Nahlik (2013) do not use the terms or make an explicit distinction between "stock" and "flow" concepts, but the stock concept is implicit in their definition of FEGS

Table 6-1. Comparison of NESCS and FEGS (continued)

	NESCS	FEGS -CS
Classification Structure	Four main classification groups: 1. **Environmental classes** divide the Earth's systems into spatial units with similar physical characteristics 2. **Ecological end-product categories** are the biophysical components of nature, either directly used by humans to produce goods and services or directly appreciated or used to yield human well-being 3. **Direct human use/non-use categories** represent the different ways in which the end-products of nature are directly used or appreciated by humans 4. **Direct human user** categories are the different sectors of the economy (including households) that directly use or appreciate the end-products. The first two groups constitute the NESCS-Supply (by whom and how services are produced). The third and fourth groups constitute the NESCS-Demand (by whom and how services are used).	Two main classification groups: 1. **Environmental classes** divide the Earth's surface into spatial units with similar physical characteristics 2. **Human beneficiary categories** are the interests of an individual (i.e., person, organization, household, or firm) that drive active or passive consumption and/or appreciation of ecosystem services resulting in impact on the interested party's welfare
Ecosystem Service Identification	Each unique combination of the four classification groups identifies a potential category of FFES.	Three Key Steps: 1. Clearly define the Environmental Class 2. Identify the Beneficiary Categories 3. For a combination of specific Beneficiary Category and Environmental Class, hypothesize FEGS received.

6.4   Other Potential Applications for NESCS

Although the primary motivation for developing NESCS is to support environmental policy analysis, we expect NESCS will provide a useful framework for other applications as well. NESCS could also potentially be used to analyze other policies (e.g., housing, transportation, tax policies) that could also result in changes to ecosystems. In Section 3.1.2 we discussed how both macro- and micro-level accounting systems can and are being adapted to address non-market elements, in particular the contributions of natural and environmental systems. Since the concepts and approach of the national economic accounts provide the underlying principles and tools for NESCS (e.g., dual-supply and demand-side classification systems), we expect that NESCS can also support efforts to expand NIPA accounts to include ecosystem services (i.e., green GDP accounting). However, it is important to keep in mind that

NIPA accounts focus on total levels of production rather than on the effects of policy changes, which are the main focus of NESCS.

NESCS can also be useful for private-sector (micro-level) environmental systems. There is a growing interest in incorporating environmental factors in private-sector corporate accounts (See Section 3.1.2 for more details). NESCS may help support alternative frameworks that are used for private-sector accounting. During the course of developing the NESCS system, we have explored whether NESCS can potentially be applied to Full Cost Accounting (FCA) frameworks adopted by the private sector. We found that at present, barring a few exceptions, accounting for ecosystem services is still not part of the FCA structure. Given the rising interest in ecosystem services, this may change in the future. There may be other micro-level accounting frameworks such as life-cycle assessments that may be relevant for NESCS.

6.5   Suggested Next Steps and Future Research

The goal of NESCS is to support policy analysis such as CBA. The key features of NESCS described in Section 6.2 will play an important role in this. It should be noted that some of these features are unique to NESCS. For example, the systematic definition of services as *flows* are an advancement over the FEGS-CS. The explicit separation of the providers and consumers of these services are an advancement over CICES. In addition, the modularity of NESCS provides more comprehensiveness and flexibility than both FEGS-CS and CICES. Separating the uses and users, rather than representing them as beneficiaries for example, provides more flexibility to an analysist conducting valuation.

The policy applications described in Section 5 demonstrate how NESCS can be applied to real-life policy questions; however, the existing structure can be further developed, refined, and expanded in several ways. Some of the possible and recommended next steps are described below.

1. *Address remaining issues and challenges to support the goal of identifying distinct pathways:*

Although NESCS provides a detailed and structured approach for identifying and classifying FFES, it does not completely resolve all of the issues associated with ecosystem service classification. Some of the key issues and remaining challenges encountered during the development of NESCS include the following topics.

144

First, for conceptual purposes the NESCS conceptual framework draws a bright line between natural systems and human systems. It depicts FFES as flows that specifically cross this line (from nature to humans). In practice, however, it is often difficult to know where to draw this line. Because some degree of human management is present in most ecosystems, it does not make sense to define natural systems so narrowly that they exclude any areas with a human footprint. But the level of human management varies widely across ecosystems, creating "gray" areas that blur the line between natural and human systems. Whereas undeveloped and sparsely populated wilderness areas can easily be thought of as natural systems, it is less clear how to characterize heavily managed natural systems such as reservoirs, agricultural systems, and sand-renourished beaches.

To help in drawing this line for NESCS, we argue that any good or service that is (1) produced by humans, and (2) intended for sale in markets is *not* an ecosystem service. In other words, for example, agricultural production and commercial harvesting of fish produce flows of *economic* goods rather than ecosystem services. In the NESCS conceptual framework, these flows occur strictly within and between human systems. For these production systems, final ecosystem services are the inputs that do not meet the two criteria above—for example, FFES from soils and precipitation to farmers and FFES from ocean fish stocks to commercial fishers.

Unfortunately, not all gray areas are resolved by these criteria. In particular, there are continuing questions about how to handle the outputs of natural systems that are heavily managed by humans, but not intended for sale in markets. For example, by planting and maintaining trees for sale, plantation forests produce economic goods (e.g., saw logs); however, they may also produce external benefits by filtering air pollutants, regulating stormwater, and providing aesthetic amenities. *In these particular roles*, it may make sense to treat the trees as inputs from natural systems rather than as outputs of market systems. Other examples are publicly owned and managed natural systems such as certain reservoirs, fish hatcheries, and renourished beaches. In these cases, public sector activities produce "natural" resources that are generally not sold in markets (although access fees may in some cases apply), but ambiguity exists regarding whether to treat these as natural systems with FFES outputs or as human production systems using natural systems as FFES inputs.

A second issue arises in trying to define ecological end-product categories that are mutually exclusive but still capture the main natural features that direct users care about. This process is challenging because different uses of a natural resource may rely on different individual attributes or different bundles of attributes. For example, as discussed in Section 4, sugar maple trees support multiple direct uses including maple syrup extraction, timber harvesting, and fall foliage aesthetics. One approach is to treat the tree as the end-product. Another approach might be to define end-products by further subcategorizing the tree into its components or attributes (wood, sap, leaves, etc.) to target the specific uses. Alternatively, it might be the case that the end-product for fall foliage viewing is not the leaves themselves, but a bundle of attributes that define the entire landscape, including other trees and natural features. In this case, using a composite end-product, such as a landscape, may correspond best with the particular use, but it will also result in overlapping end-product categories (i.e., sugar maple trees included as individual end-products and as components of a landscape end-product). However, these types of overlaps may not be a problem, as long as they do not lead to double counting of certain FFES.

A third issue is whether or how to address differences or changes in the quality characteristics of ecological end-product in NESCS. For example, when a policy action leads to improvements in water quality by increasing water clarity, where will these changes and the resulting increases in ecosystem services be captured in NESCS or in applications of its classification structure? We argue that the best way to address quality differences would be through the quantification and valuation of ecosystem services rather than through the classification system itself. For example, just as quality differences between market goods and services are reflected in their market prices, quality differences in ecological end-products can be captured through differences in their estimated values. Meanwhile, quality indicators, such as water clarity and frequency of algal blooms, can be used to represent the end-product attributes that people value, but these indicators would not be used to define mutually exclusive categories within the classification system.[75]

A fourth issue is who to define as the direct user of an ecological end-product, when access to in-situ use of the end-product is provided through a market transaction. For example, many ecotourism services, such as chartered fishing excursions or nature tours, provide customers with access to specific ecological end-products, such as ocean fish stocks or areas of

---

[75] In principle, one potential alternative way to address quality changes would be to subdivide the end-products into quality-related subcategories. Just as NAICS/NAPCS have separate categories for first- and second-class air travel, the water end-product could in principle be divided into low- and high-quality subcategories (i.e., water quality changes would be captured by differences in the number of waterbodies in each subcategory). However, this approach is unlikely to be feasible on a large scale and is not recommended. The number of these types of subcategories would need to be limited to avoid overcomplicating NESCS, and the criteria for assigning waters to the different categories would need to be carefully considered. Moreover, assigning a quality level to an ecological end-product requires consideration of how the end-product is used by humans (e.g., water quality criteria for fishing and swimming can be very different). In other words, it would require mixing NESCS-D concepts into the NESCS-S classification.

natural beauty. Similarly, home builders receive a premium in the market by building and selling structures with natural vistas. In these cases, are the sellers or buyers (or both) the direct users? One perspective is that, in these cases, the end-products are inputs to the production process; therefore, the sellers are the direct users. The buyers benefit but only indirectly through the sellers. Another perspective is that the buyers are the direct users because they are the ones who directly experience and enjoy the end-product. The sellers benefit but only indirectly through payment from the buyers. In our policy examples in Section 5, we have allowed for the two perspectives in different contexts. For example, in Table 5-10, households are direct users of clean water through scenic amenities from waterside homes, and charter fishing trip providers are direct users of fish stocks to provide recreational trips. Arguments can be made for both perspectives, as long as they are not used together in a way that results in double counting of FFES.

2. Make explicit linkages to other classification and accounting systems:

Developing a cross-walk between four-group NESCS categories and two-group FEGS-CS categories would be a useful next step. As described in Table 6-1, since the goal of the FEGS-CS is to determine those specific ecosystem attributes that the beneficiary values, this can help identify appropriate metrics and indicators. Thus, it can complement the NESCS structure in a way that will further support policy analysis. Links and complementarities with the other ecosystem service classification and accounting systems (such as SEEA-EEA) described in Section 2 will be explored.

3. Test and evaluate NESCS through additional applications:

Continuing to evaluate and demonstrate NESCS using additional policy application examples will be important to further refine NESCS. These policy applications would include environmental contexts as well as other contexts. This would also help identify other potential users of NESCS, and help expand inter-disciplinary and inter-agency collaborations.

SECTION 7
REFERENCES

Anderson, J.R., E.E. Hardy, J.T. Roach, and R.E. Witmer. 1976. *A Land Use and Land Cover Classification System for Use with Remote Sensor Data*. Geological Survey Professional Paper 964. Washington, DC: U.S. Government Printing Office.

Berrittella, M., A. Bigano, R. Roson, and R.S.J. Tol. 2006. A general equilibrium analysis of climate change impacts on tourism. *Tourism Management* 27(5): 913–924.

Bosello, F., R. Roson, and R.S.J. Tol. 2007. Economy-wide estimates of the implications of climate change: Sea level rise. *Environmental and Resource Economics* 37(3):549–571.

Bureau of Economic Analysis (BEA). 2007. Conceptual basis of the accounts. *Measuring the Economy: A Primer on GDP and the National Income and Product Accounts*. Bureau of Economic Analysis, Department of Commerce. http://www.bea.gov/national/pdf/nipa_primer.pdf

Bureau of Economic Analysis (BEA). 2011. *Measuring the Nation's Economy: An Industry Perspective*. Bureau of Economic Analysis, Department of Commerce. https://www.bea.gov/industry/pdf/industry_primer.pdf

Boyd, J. 2007. Nonmarket benefits of nature: What should be counted in green GDP? *Ecological Economics* 61(4):716–723.

Boyd, J., and S. Banzhaf. 2007. What are ecosystem services? The need for standardized environmental accounting units. *Ecological Economics* 63:616–626.

Bureau of Economic Analysis (BEA). 2009. *Concepts and Methods of the U.S. Input-Output Accounts*. Washington, DC: U.S. Department of Commerce.

Carbone, J.C., and V.K. Smith. 2013. Valuing nature in a general equilibrium. *Journal of Environmental Economics and Management* 66(1):72–89.

Costanza, R. 2008a. Ecosystem services: Multiple classification systems are needed. *Biological Conservation* 141:350–352.

Constanza, R. 2008b. Natural capital. *The Encyclopedia of Earth*. http://www.eoearth.org/view/article/154791/

Costanza, R., R. D'Arge, R.S. de Groot, S. Farber, M. Grasso, B. Hannon, K. Limburg, S. Naeem, R.V. O'Neill, J. Paruelo, R.G. Raskin, P. Sutton, and M. van den Belt. 1997. The value of the world's ecosystem services and natural capital. *Nature* 387(6630):253–260.

Daily, G.C., S. Alexander, P.R. Ehrlich, L. Goulder, J. Lubchenco, P.A. Matson, H.A. Mooney, S. Postel, S.H. Schneider, D. Tilman, and G.M. Woodwell. 1997. Ecosystem services: Benefits supplied to human societies by natural ecosystems. *Issues in Ecology* 2:1–16.

149

Daily, G. 1997. *Nature's Services: Societal Dependence on Natural Ecosystems*. Washington, DC: Island Press.

de Groot, R., B. Fisher, M. Christie, J. Aronson, L. Braat, R.H. Haines-Young, J. Gowdy, T. Killeen, E. Maltby, A. Neuville, S. Polasky, R. Portela, and I. Ring. 2010. Integrating the ecological and economic dimensions in biodiversity and ecosystem service valuation. Draft Chapter 1 of *The Economics of Ecosystems and Biodiversity (TEEB) Study*.

de Groot, R., M. Wilson, and R. Boumans. 2002. A typology for the classification, description and valuation of ecosystem functions, goods and services. *Ecological Economics* 41:393–408.

Economic Classification Policy Committee (ECPC). 1993a. *Issues Paper No. 1, Conceptual Issues*. http://www.census.gov/eos/www/naics/history/history.html

Economic Classification Policy Committee (ECPC). 1993b. *Issues Paper No. 2, Aggregation Structures and Hierarchies*. http://www.census.gov/eos/www/naics/history/history.html

Economic Classification Policy Committee (ECPC). 1993c. *Issues Paper No. 3, Collectability of Data*. http://www.census.gov/eos/www/naics/history/history.html

Economic Classification Policy Committee (ECPC). 1993d. *Issues Paper No. 4, Criteria for Determining Industries*. http://www.census.gov/eos/www/naics/history/history.html

Economic Classification Policy Committee (ECPC). 1993e. *Issues Paper No. 6, Service Classifications*. http://www.census.gov/eos/www/naics/history/history.html

Economic Classification Policy Committee (ECPC). 1993f. *Public Comments on Issues Papers No. 1 & 2*. http://www.census.gov/eos/www/naics/history/history.html

Economic Classification Policy Committee (ECPC). 1994. *ECPC Report 1, Economic Concepts in SIC Industries*. http://www.census.gov/eos/www/naics/history/history.html

Economic Classification Policy Committee (ECPC). 2001. *Accounting for Non-Market Products and Activities in NAPCS*. http://www.census.gov/eos/www/napcs/discussionpapers.html

Economic Classification Policy Committee (ECPC). 2003. *Overview of NAPCS Objectives, Guidance, and Implementation Strategy and Goals: A United States Perspective*. http://www.census.gov/eos/www/napcs/discussionpapers.html

Fisher, B, and R.K. Turner. 2008. Ecosystem service: Classification for valuation. *Biological Conservation* 141:1167–1169.

Fisher, B., R.K. Turner, and P. Morling. 2009. Defining and classifying ecosystem services for decision making. *Ecological Economics* 68(3):643–653.

Fu, B., C. Su, Y. Wei, I. Willet, Y. Lu, and G. Liu. 2011. Double counting in ecosystem services valuation: Causes and countermeasures. *Ecological Research* 26:1–14.

Ghimire, S.K. and Y. Aumeeruddy-Thomas. 2009. Ethnobotanical classification and plant nomenclature system of high altitude agro-pastoralists in Dolpo, Nepal. *Journal of Plant Science* 6:56–68.

Haines-Young, R., and M. Potschin. 2013. *Common International Classification of Ecosystem Services (CICES): Consultation on Version 4, August-December 2012*. Report to the European Environment Agency. EEA Framework Contract No EEA/IEA/09/003.

Haines-Young, R., and M. Potschin. 2010a. "The links between biodiversity, ecosystem services and human well-being." *Ecosystem Ecology: A New Synthesis. BES Ecological Reviews Series*. Cambridge: Cambridge University Press.

Haines-Young, R., and M. Potschin. 2010b. *Proposal for a Common International Classification of Ecosystem Goods and Services (CICES) for Integrated Environmental and Economic Accounting*. http://www.nottingham.ac.uk/cem/pdf/UNCEEA-5-7-Bk1.pdf

Hanson, C., J. Ranganathan, C. Iceland, and J. Finisdore. 2012. *The Corporate Ecosystem Services Review*. World Resources Institute. http://www.wri.org/publication/corporate-ecosystem-services-review

Hawkins, K. 2003. *Economic Valuation of Ecosystem Services*. University of Minnesota. http://mn.gov/frc/documents/council/landscape/SE%20Landscape/MFRC_Economic_Valuation_EcosystemServices_SE_2003-10-01_Report.pdf

Hein, L., K. van Koppen, R.S. de Groot, and E.C. van Ierland. 2006. Spatial scales, stakeholders and the valuation of ecosystem services. *Ecological Economics* 57:209–228.

Herrendorf, B., R. Rogerson, and A. Valentinyi. 2013. Two perspectives on preferences and structural transformation. *American Economic Review* 103:2752–2789.

International Petroleum Industry Environmental Conservation Association [IPIECA]. 2011. *Ecosystem Services Guidance: Biodiversity and Ecosystem Services Guide and Checklists*. International Association of Oil and Gas Producers (OGP) Report No. 461. London, United Kingdom. http://www.ipieca.org/publication/ecosystem-services-guidance-biodiversity-and-ecosystem-services-guide

Johnston, R.J., and M. Russell. 2011. An operational structure for clarity in ecosystem service values. *Ecological Economics* 70(12):2243–2249.

Lancaster, Kelvin. 1966. Change and innovation in the technology of consumption. *The American Economic Review* 14–23.

Landers, D.H., and A.M. Nahlik. 2013. *Final Ecosystem Goods and Services Classification System* (FEGS-CS). EPA/600/R-13/ORD-004914. Washington, DC: U.S. Environmental Protection Agency, Office of Research and Development.

Lindenmayer, D.B. and R.B. Cunningham. 1996. A habitat-based microscale forest classification system for zoning wood production to conserve a rare species threatened by logging operations in south-eastern Australia. *Environmental Monitoring and Assessment* 39:543–557.

Maxwell, D., E. McKenzie, and R. Traldi. 2014. *Valuing Natural Capital in Business: Towards a Harmonised Protocol*. Natural Capital Coalition. http://www.naturalcapitalcoalition.org/js/plugins/filemanager/files/Valuing_Nature_in_Business_Part_1_Framework_WEB.pdf

Mead, C.I., K. Moses, and B.R. Moulton. 2004. The NIPAs and the System of National Accounts. *Survey of Current Business* 84:17–32.

Millennium Ecosystem Assessment (MA). 2003. *Ecosystems and Human Well-Being: A Framework for Assessment*. Island Press, Washington, DC.

Millennium Ecosystem Assessment (MA). 2005. *Ecosystems and Human Well-Being: Synthesis*. Island Press, Washington, DC.

Moberg, F. and C. Folke, 1999. Ecological goods and services of coral reef ecosystems. *Ecological Economics* 29(2):215–233.

Mohr, M.F. 2002. *North American Product Classification System (NAPCS): What's been done, what's next*. NAPCS Discussion Paper prepared for Census Advisory Committee of Professional Associations Meeting, April 18-19. http://www.census.gov/eos/www/napcs/papers/cacapr02.pdf

Nahlik, A.M., M.E. Kentula, M.S. Fennessy, and D.H. Landers. 2012a. Where is the consensus? A proposed foundation for moving ecosystem service concepts into practice. *Ecological Economics* 77:27–35.

Nahlik, A.M., D.H. Landers, P.L. Ringold, and M.A. Weber. 2012b. Protecting our environmental wealth: Connecting ecosystem goods and services to human well-being. *National Wetlands Newsletter* 34: 14-18.

National Ecosystem Services Partnership. 2014. Different uses of the Ecosystem Services Framework: Scenario analysis and "Green Accounting. In *Federal Resource Management and Ecosystem Services Guidebook*. National Ecosystem Partnership, Duke University, Durham, NC. https://nespguidebook.com/assessment-framework/comparison-to-green-accounting/

National Research Council. 2005. *Valuing Ecosystem Services: Toward Better Environmental Decision-Making*. Washington, DC: National Academies Press.

Norberg, J. 1999. Linking nature's services to ecosystems: Some general ecological concepts. *Ecological Economics* 29(2):183–202.

Nordhaus, W.D. 2000. New directions in national economic accounting. *The American Economic Review* 90(2):259–263.

Nordhaus, W.D. 2004. *Principles of National Accounting for Non-market Accounts*. Working Paper prepared for CRIW Conference on the Architecture for the National Accounts. http://www.nber.org/CRIW/CRIWs04/nordhaus.pdf

Ott, W., and C. Staub. 2009. *Welfare-Significant Environmental Indicators. A Feasibility Study on Providing a Statistical Basis for the Resources Policy. Summary.* Environmental Studies No. 0913. Bern: Federal Office for the Environment. http://www.environment-switzerland.ch/uw-0913-e

Porritt, J. 2007. *Capitalism as if the World Matters*. London: Earthscan.

Ramsar. 2011. *Wetland Ecosystem Services—An Introduction*. Switzerland: Ramsar Convention on Wetlands. http://ramsar.rgis.ch/pdf/info/services_00_e.pdf

Ringold, P.L., J. Boyd, D. Landers, and M. Weber. 2009. *Report from the Workshop on Indicators of Final Ecosystem Services for Streams*. Meeting July 13-16. EPA/600/R-09/137.

Ringold, P.L., J.W. Boyd, D.H. Landers, and M.A. Weber. 2013. What data should we collect? A framework for identifying indicators of ecosystem contributions to human well-being. *Frontiers in Ecology and the Environment* 11:98–105.

Ringold, P.L., A.M. Nahlik, J.W. Boyd, and D. Bernard. 2011. *Report from the Workshop on Indicators of Final Ecosystem Goods and Services for Wetlands and Estuaries*. EPA/600/X-11/014. Washington, DC: U.S. Environmental Protection Agency.

Sokal, R. 1974. Classification: Purposes, principles, progress, prospects. *Science* 185(4157):1115–1123.

Staub, C., W. Ott, F. Heusi, G. Klingler, A. Jenny, M. Häckl, A. Hauser. 2011 *Indicators for Ecosystem Goods and Services: Framework, Methodology and Recommendations for a Welfare-Related Environmental Reporting*. Federal Office for the Environment, Bern. Environmental Studies No. 1102: 17S. http://www.environment-switzerland.ch/uw-1102-e

Turner, R.K., D.W. Pearce, and I. Bateman. 1994. *Environmental Economics. An Elementary Introduction*. NY: Harvester Wheatsheaf.

United Nations (U.N.). 2003. *Handbook of National Accounting: Integrated Environmental and Economic Accounting 2003*. http://unstats.un.org/unsd/environment/seea2003.pdf

U.N. Department of Economic and Social Affairs, 1999. *UN Glossary of Classification Terms (Short Version)*. ESA/STAT/AC.75/8. http://unstats.un.org/unsd/class/intercop/expertgroup/1999/ac75-8a.pdf

U.N. Department of Economic and Social Affairs. 2012. *System of Environmental Economic Accounting for Water*. ST/ESA/STAT/SER.F/100. http://unstats.un.org/unsd/envaccounting/seeaw/seeawaterwebversion.pdf

U.S. Environmental Protection Agency. (2009). *Valuing the Protection of Ecological Systems and Services: A Report of the EPA Science Advisory Board*. Report EPA-SAB-09-012. Washington, DC: U.S. Environmental Protection Agency.

Waage, S. and C. Kester. 2013. *Private Sector Uptake of Ecosystem Services Concepts and Frameworks: The Current State of Play*. BSR Report. http://www.bsr.org/reports/BSR_Private_Sector_Uptake_Ecosystem_Services.pdf

Wainger, L.A., and M.J. Mazzotta. 2011. Realizing the potential of ecosystem services: A framework for relating ecological changes to economic benefits. *Environmental Management* 48:710–733. http://www.springerlink.com/content/l59666mm22777225/

Wallace, K.J. 2007. Classification of ecosystem services: Problems and solutions. *Ecological Conservation* 139:235–246.

Wallace, K.J. 2008. Ecosystem services: Multiple classifications or confusion? *Ecological Conservation* 141:353–354.

WAVES (Global Partnership on Wealth Accounting and the Valuation of Ecosystem Services). 2013. *Annual Report*. WAVES. https://www.wavespartnership.org/sites/waves/files/images/WAVES-Annual-Report.pdf

Wu, J., and T. Wu. 2010. Green GDP in *Berkshire Encyclopedia of Sustainability, Vol. II–The Business of Sustainability*. Great Barrington: Berkshire Publishing.

GLOSSARY OF KEY TERMS USED IN THE NESCS REPORT

Term	Definition
Classification system	Provides an organized structure, through well-defined categories that allow one to group similar elements together and to separate others. Pre-determined criteria define what should be considered similar or different, and these criteria are driven by the specific purpose for developing the classification system.
Direct use/non-use	Different ways in which end-products are directly used or appreciated by humans. Direct uses may be either extractive or in-situ. End-products may be used as inputs into market production processes or they may be used or appreciated by households. Note that households may derive well-being from actually using end-products as well as from non-use (i.e., households may appreciate end-products even if they do not see or use them).
Direct users	Sectors of society/economy that directly use or appreciate the end-products
Ecosystems	The dynamic complex of plant, animal, microorganism communities, and the non-living environment which together interact as a system
Ecosystem services	The ways in which ecosystems contribute to human well-being
Ecological end-products	Biophysical components of nature that are either directly used by humans to produce goods and services or directly enjoyed or used to yield human well-being. They are usually (but not always) represented as stocks of end-products. Note that conceptually, they are different from FFES (defined below) but in some situations may be used as indicators of FFES
Environment	Spatial units, with similar biophysical characteristics, that are located on or near the Earth's surface and that contain or produce "end-products"
Final Ecosystem Goods and Services (FEGS)	Components of nature, directly enjoyed, consumed or used to yield human well-being (The concept of "final" ecosystem services was developed by Boyd and Banzhaf (2007) and Landers and Nahlik (2013) adopted the term FEGS later to represent this concept.)
Final Ecosystem Goods and Services Classification System (FEGS-CS)	A two-group classification system developed by Landers and Nahlik (2013). FEGS are identified by the landscape in which they occur (Environmental Class) and the interests of the people that interact with the FEGS (Beneficiary Categories).
Flows of Final Ecosystem Services (FFES)	The contributions that the end-products of nature provide (1) directly to human production processes or (2) directly to households and human well-being. They are represented by service flows between ecological end-products and direct human uses. Note that conceptually, they are different from end-products (defined above).
Final economic services	Final economic services are sold to the end user i.e., flow from producers to households.
Flows	A flow variable is measured over an interval of time. Therefore, flow measures are typically expressed as a rate per unit of time—e.g., annual income (dollars/year) and daily nutrient load (pounds per day).
Marginal Analysis	Analysis of policies that involves evaluations of *changes* to the system rather than evaluating the status of the *total* system. Policies that are relevant in this context are typically those that cause *changes* to ecosystems that are small *relative* to the *total* value of ecosystems.

Term	Definition
National Ecosystem Services Classification System (NESCS)	A classification system for flows of final ecosystem services. It provides a conceptual framework, a four-group classification structure, and a coding system for identifying distinct FFES. It is designed primarily to support the analysis of welfare impacts of policy-induced changes to ecosystems. Note that the NESCS terminology does not include flows of final ecosystem goods. NESCS defines (1) ecological end-products (most of which are stocks of ecosystem goods), (2) flows of final ecosystem services, and (3) flows of economic goods.
Natural capital	Natural capital is the stock of natural ecosystems that yields a flow of valuable ecosystem goods or services into the future (Costanza, 2008b). In the context of NESCS, it is important to consider both quantity as well as quality attributes of natural capital. This is because changes in policy can lead to changes in one or both of these attributes and consequently lead to changes in the FFES provided.
NESCS-D	Demand-side classification in NESCS that characterizes how and by whom FFES are used/appreciated and consists of two groups: Direct Use/Non-Use and Direct Users
NESCS-S	Supply-side classification in NESCS that characterizes how and by whom FFES are provided and consists of two groups: Environment and End-Products
Non-market valuation	Methods used to estimate values of goods and services that are typically not exchanged in markets and therefore do not have associated observable transactions.
Non-use values	Human preferences for goods or services that are not associated with or derived from direct use or contact with them. For instance, individuals may care about or appreciate ecological end-products, even if they never directly use or see them – i.e., they may have non-use values for the existence of things like tropical forests or pristine lakes, even if they never visit them. They are distinct from "use" values.
Use values	Human preferences for goods or services that are associated with or derived from direct use or contact with them.
Services	Services are distinct from goods. Services are typically intangible, non-storable, and inseparable from provider and consumer. Also, typically in economics, in contrast to goods, which can be treated as "stocks" and measured at a specific point in time, services are viewed as "flows" from the provider to the consumer and are measured over a period of time.
Stock	A stock variable represents a quantity existing at a point in time (which may have accumulated in the past). Units of measurement are typically expressed in levels – e.g., wealth (dollars), physical assets (number of machines), and nutrient concentration (milligrams per liter) at the beginning of the year.
Total Economic Value (TEV) Framework	Broad conceptual framework commonly used by economists to organize different types of values (e.g., use and non-use values) that may be associated with a good or service. See chapter 4 for an example of a commonly used TEV framework.

APPENDIX A
MATHEMATICAL REPRESENTATION OF THE CONCEPTUAL MODEL

The purpose of this appendix is to provide a more formal and mathematical representation of the conceptual framework for linking a policy action and its impact on ecosystems to resulting changes in human well-being. In this framework a policy action (ΔZ) is assumed to cause changes to natural systems (ΔN), which then leads to changes in the ecological end-products (ΔE) that are directly used or appreciated by humans. Changes in ecosystem productivity and in the profile of end-products (ΔE resulting from ΔN) can take many different forms, but in each case the result is to alter the flows of final ecosystem services (FFES) to humans, either by altering the production of the final economic goods and services they consume (ΔY, path 1 from ΔE) which then affects their well-being (ΔW), or by directly affecting their well-being (by way of the curved arrow, path 2 from ΔE)

$$\Delta Z \rightarrow \Delta N \rightarrow \Delta E \xrightarrow{1} \Delta Y \rightarrow \Delta W$$

Generally speaking, the benefits of a policy-related change can be represented by the marginal utility/well-being with respect to this policy change.

$$\frac{\partial W}{\partial Z} = \left(\frac{\partial N}{\partial Z} * \frac{\partial E}{\partial N}\right) * \left(\frac{\partial Y}{\partial E} * \frac{\partial W}{\partial Y} + \frac{\partial W}{\partial E}\right) \tag{A.1}$$

In any "marginal" analysis, it is assumed that the changes in Z, N, E, and Y are relatively small compared to the total economy and to all ecosystems; however, they still have a meaningful effect on human well-being. In addition, N, E, Y, and W are all vectors, implying that there are multiple avenues through which the policy change can affect human well-being (as shown in Figure 4-5.

The components of equation (A1.1) can be described as follows:

- The first term on the right-hand side — $\frac{\partial N}{\partial Z}$ — represents the marginal direct impact on the quantity and/or quality of natural systems (*N*) with respect to the change in policy (*Z*). For example, it could be the additional number of tidal wetland acres protected from destruction by a coastal management policy.

- The second term on the right side — $\frac{\partial E}{\partial N}$ — natural systems, or natural capital, with respect to end-products – i.e., the additional amount of ecological end-product generated per additional unit of N. For example, this could include the increase in striped bass populations resulting from the additional protected wetland acres.

- The third term — $\frac{\partial Y}{\partial E}$ — is the marginal product of the market production function with respect to a change in the profile of ecological end-products (i.e., the additional output of final economic goods and services generated per additional unit of input E). This term reflects and is an indicator of the increase in final ecosystem service flows (FFES).[76] For example, this could represent the increment in fish supplied to the market (holding all other production inputs constant, such as labor and capital inputs) as a result of larger fish stocks in the wild.

- The fourth term — $\frac{\partial W}{\partial Y}$ — is the marginal utility (i.e., marginal well-being) per additional unit of final market goods and services produced. In dollar terms, it can be interpreted as the marginal value (i.e., price) per unit of additional output in the market Y (e.g., the price of fish).

- The final term — $\frac{\partial W}{\partial E}$ — is the marginal utility directly experienced by households per additional unit of non-marketed characteristics of E. For example, it could include the increment in utility from recreational fishing associated with a unit increase in the striped bass fish population. It reflects the increase in FFES directly flowing to households and can be interpreted as the non-market value (implicit price) per additional unit of E.

In this formulation, the total benefit of a policy-induced change in an ecological end-product (E) is equal to the sum of impacts on well-being experienced (1) indirectly through changes in inputs to market production processes, and (2) through direct, non-market-related changes in human well-being.

In this formulation, the FFES are primarily captured in the terms $\frac{\partial Y}{\partial E}$ and $\frac{\partial W}{\partial E}$, which represent the marginal product and marginal utility of the ecological end-products. In other words, the presence of an FFES requires that $\frac{\partial Y}{\partial E} > 0$ and/or $\frac{\partial W}{\partial E} > 0$. If a change in an ecological end-product does not increase market output and/or human well-being (holding the flow of all other goods and services constant), then it does not provide an ecosystem service.[77]

---

[76] Using the marginal product of E as an indicator of FFES is similar to using the marginal product of labor or capital to represent the services they provide to producers.

[77] Negative effects would imply ecosystem "disservices," which are also possible (e.g., nuisance effects of mosquitos from wetlands).

APPENDIX B
EXPANDED CONCEPTUAL FRAMEWORK FOR
ECOSYSTEM SERVICES ANALYSIS

This appendix expands on the conceptual framework described in Section 4 (specifically Figure 4-3) by describing in more detail the connections between natural capital and human well-being. This expanded framework does not change how ecosystem services are defined or classified in NESCS. It does, however, provide a more comprehensive representation of the input-output relationships within and between natural and humans systems. Understanding these relationships is vital for applying NESCS, and for comprehensively quantifying and valuing the effects of policy-induced changes to natural capital through to human well-being.

B.1   Intermediate Ecological Production and Ecosystem Services

In Section 4, to simplify the representation of ecological production processes, Figure 4-3 does not distinguish between intermediate and final ecological production in the same way that it separates intermediate and final economic production. However, it is important to acknowledge that the input-output relationships between ecological production systems can be as or more complex than those between economic production systems. Therefore, Figure B-1 expands the conceptual diagram to explicitly show the parallel intermediate-to-final production processes in natural and human systems.

Intermediate ecological production represents the multitude of natural processes that generate output flows that contribute indirectly to human well-being but are not directly used or appreciated by humans. For example, the processes underlying the MA concept of "supporting" ecosystem services, such as nutrient cycling, primary production, and soil formation, can for the most part be thought of as intermediate ecological production processes. As in economic production systems, the assignment of intermediate or final production depends on the context— in other words, what is final in one context can be intermediate in another. For example, the wetland process of filtering sediment from surface water can be conceptualized as a final ecological production process when it produces water clarity that is directly appreciated by humans. The same process can also improve habitat for benthic biota in streams. This type of output is typically not directly appreciated or used by humans but can be vital to the food chain that supports highly valued recreational fisheries.

Figure B-1. Expanded Conceptual Framework with Intermediated Ecological Production and Ecosystem Services

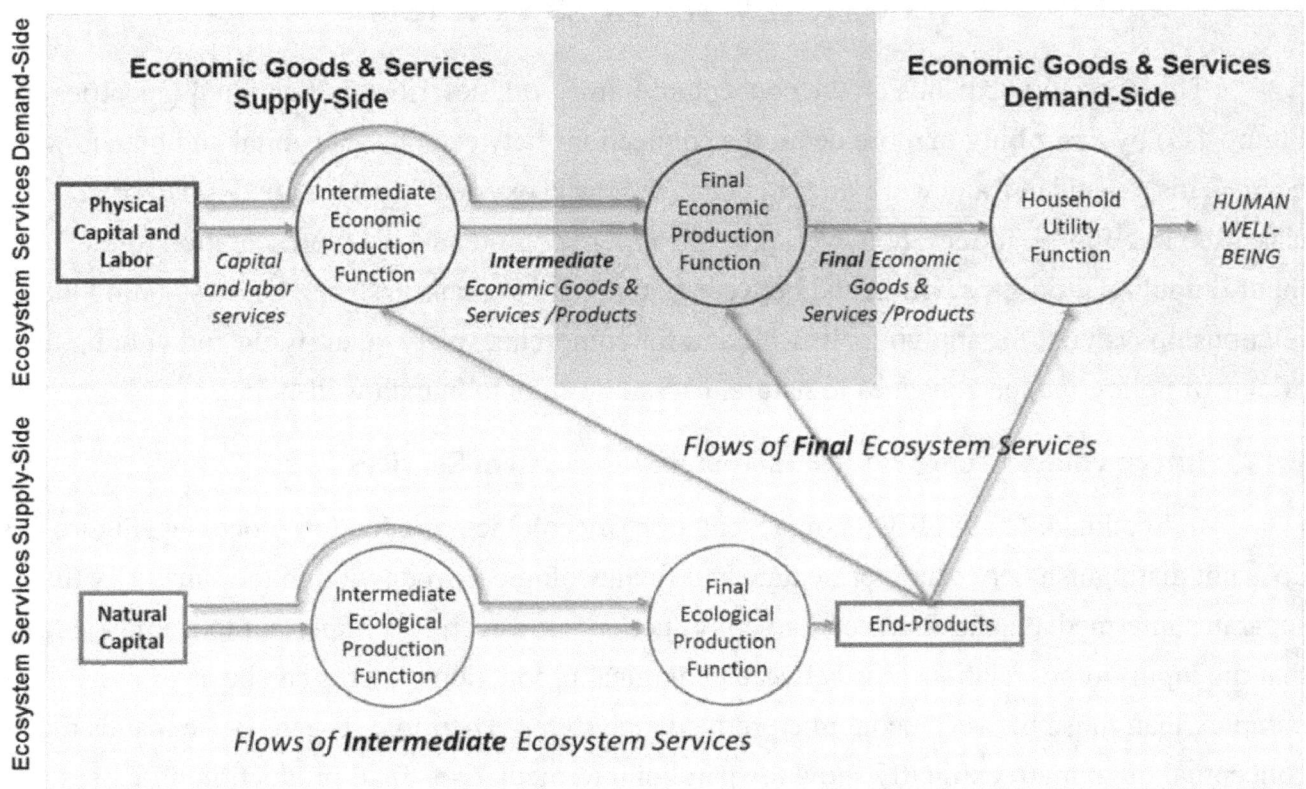

B.2   Feedbacks, Spillovers, and General Equilibrium Effects

The primary purpose of Figure 4-3 is to provide a diagrammatic representation of FFES and where they occur along the continuum from natural capital to human well-being. However, the connections between and among natural and human systems are considerably more complex than those represented in Figure 4-3. Therefore, when assessing the impacts of policies that affect natural capital, it is important to consider not only the flows shown in this figure, but also a range of other linkages that are not shown in the figure.

As previously noted, recognizing these additional connections does not require an alteration of the NESCS structure (i.e., the categories of flows between End-Products and Direct Use/Users); however, it does change how the NESCS structure is applied. In particular, depending on the context, it may require broadening (1) the spatial scale of the analysis to include geographic areas that are indirectly affected by policy actions, or (2) the temporal scale of the analysis to include dynamic feedback effects into the future.

In this appendix, we emphasize two main types of additional connections:

1. Feedbacks from human systems to natural systems;

2. Spillover effects within economic systems and within natural systems.

Figure B-2 expands Figure B-1 to show feedbacks from human systems to natural systems. For completeness, it also expands the household utility function circle to include "household production." This latter addition recognizes that households engage in non-market production activities that transform inputs from economic and natural systems into goods and services that they value. For example, households can be thought of as using transportation, recreational equipment purchases, and their own time to produce recreation trips. Like the previously included economic production functions, these activities can have both positive and negative effects on natural systems.

Figure B-2. Expanded Conceptual Framework Showing Feedback Effects from Human to Natural Systems

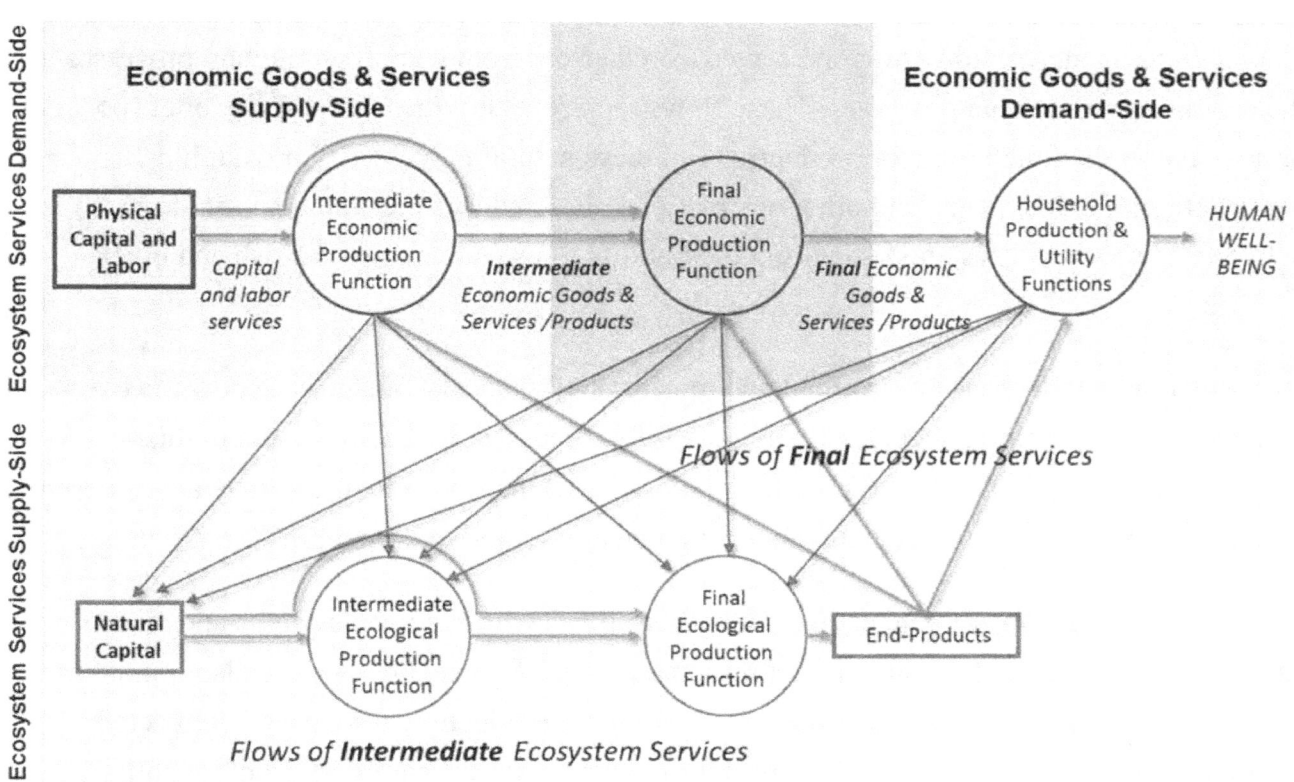

In Figure B-2, several feedback effects are represented as red arrows flowing from the economic and household production processes to natural capital. It is important to first note that these feedback flows can have both positive and negative effects on natural capital. They include:

- Depletion of natural capital stocks as a result of consumptive market and non-market human activities.[78] For example, commercial fishing activities that decrease ocean fish stocks and private well use by households that decrease groundwater stocks.

- Degradation of the quality of natural capital as a result of market and non-market human activities. For example, pollutant discharges to water from commercial establishments and households.

- Remediation and restoration of natural capital. For example, clean-up of waste sites and stream restoration projects.

- Development of urban parks and greenways.

Other feedback flows are represented as red arrow from human production processes to the ecological production processes. These flows represent activities that directly alter (positively or negatively) the functioning or productivity of these natural processes. For example, the construction of roads, dams, and other obstructions often reduces the connectivity of stream networks and wetland systems, impairing their ability to provide habitat for fish and other wildlife.

These feedback effects from human to natural systems also underscore how the two systems are often interdependent and integrated. As discussed in Sections 4 and 6, this interconnectivity can make it difficult to strictly separate the two systems, as they are represented in these figures (i.e., by the separate blue and green areas), and to define FFES in a standard way.

The second type of connection—spillover effects within human systems and within natural systems—are not specifically represented in these figures, but they are also important to consider and account for. On the human side, economic production systems are connected through markets and input-output relationships (as described in Section 3). Consequently, policy-induced changes in production activities in one economic sector can have ripple effects through other sectors, primarily through changes in relative prices. These types of market-based

---

[78] Depletion implies that rates of extraction by humans exceed the natural capital's ability to renew or replenish itself.

economic spill-over effects are typically captured using computable general equilibrium (CGE) models. For example, Berrittella et al. (2006) and Bosello et al. (2007) examine how climate change impacts the tourism sector and coastal economies (respectively), as well as how these impacts have economy-wide effects.

Even when non-market systems (e.g., households) are the entities directly affected by changes in natural capital and FFES, their connections to the broader economy can also have ripple effects in the market. These spillovers into other parts of the economy eventually feed back into households' well-being. For example, Carbone and Smith (2013) use a case study of nitrogen and sulfur emissions control policies in the United States to show how ecosystem services related to non-market uses, such as recreational fishing, aesthetic enjoyment of forest scenery, and non-use values, can be formally incorporated into a CGE framework. They do this in part by specifying a utility function that explicitly accounts for the link (non-separability) between households' preferences for market goods and for non-market ecosystem services.

On the side of natural systems, there are also innumerable connections between ecological production processes, which would ideally be accounted for in a comprehensive assessment of policy impacts. Figures B-1 and B-2 represent these processes in a linear and sequential way; however, the input-output connections between these processes are likely to be much more complex. For example, policies that reduce sediment loads, increase water clarity, and help restore seagrass beds in an estuary can improve habitat for certain crab species. However, if these crab species are also oyster predators, they may indirectly impair water clarity by limiting the water filtration function performed by oysters. Understanding these potential feedback effects is essential for fully identifying and quantifying the changes in ecological End-Products, which are the source of FFES.

Additional Information

This document was developed under U.S. EPA Contract EP-W-11-029 with RTI International (Paramita Sinha and George Van Houtven), in collaboration with the ORISE Participant Program between U.S. EPA and U.S. DOE (Charles R. Rhodes), under the direction of Joel Corona and Dixon Landers, U.S. EPA, Office of Water and Office of Research and Development. Peer review for this report was conducted under U.S. EPA Contract EP-C-12-045 with Versar, Inc. (David Bottimore).

This report may not necessarily reflect the views of U.S. EPA and no official endorsement should be inferred.

To provide feedback on this report or any other aspect of the NESCS approach, please send comments by email to NESCS@epa.gov.

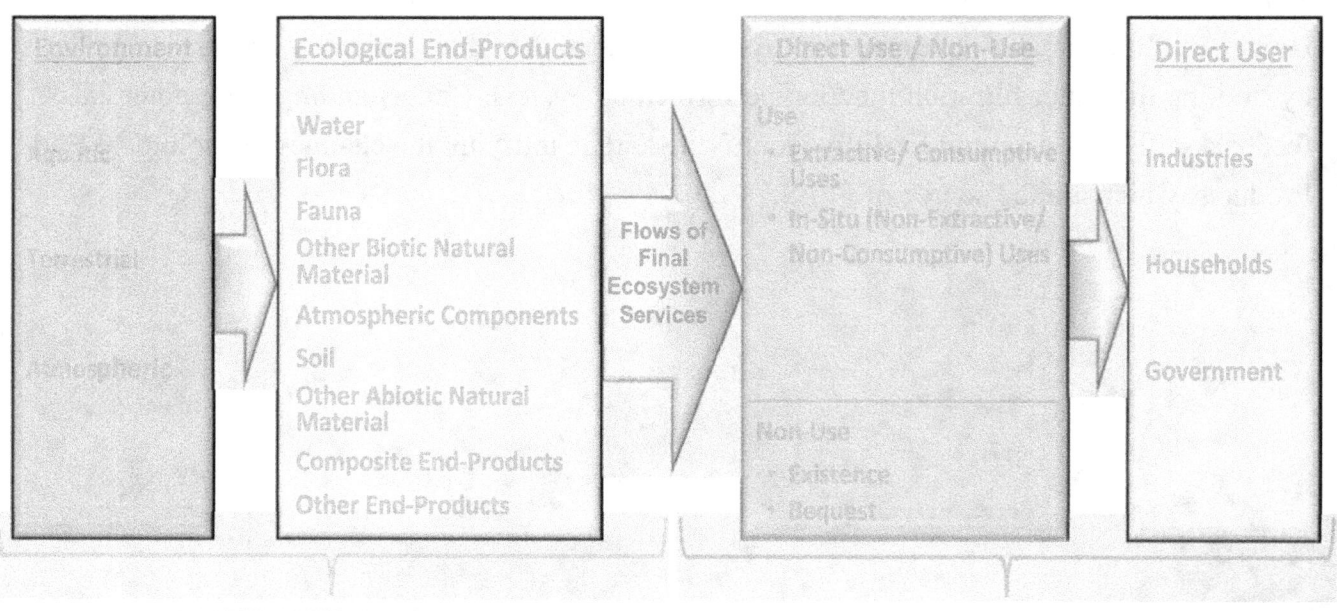

United States Environmental Protection Agency. 2015. *National Ecosystem Services Classification System (NESCS): Framework Design and Policy Application.* EPA-800-R-15-002. United States Environmental Protection Agency, Washington, DC.

EPA-800-R-15-002
September 2015